How to Love a Rat

ATELIER: ETHNOGRAPHIC INQUIRY IN THE TWENTY-FIRST CENTURY

Kevin Lewis O'Neill, Series Editor

How to Love a Rat

DETECTING BOMBS IN POSTWAR
CAMBODIA

Darcie DeAngelo

UNIVERSITY OF CALIFORNIA PRESS

University of California Press
Oakland, California

© 2024 by Darcie Anne Hoffman DeAngelo

Cataloging-in-Publication data is on file at the Library of Congress.

ISBN 978-0-520-39740-8 (cloth)

ISBN 978-0-520-39742-2 (pbk.)

ISBN 978-0-520-39743-9 (ebook)

33 32 31 30 29 28 27 26 25 24
10 9 8 7 6 5 4 3 2 1

To my parents

Contents

Acknowledgments

This book emerged after years of work that depended on so many people. Anthropology is not a solo project, despite its long history of pretending to be. I first want to thank the platoon of deminers who welcomed me into their workspace and who allowed me to watch them learn something new, an always vulnerable time. They were beyond gracious and generous. I also want to thank the staff and leaders at Cambodian Mine Action Center and the NGO with the landmine detection rats, APOPO. They allowed me to witness a landmine detection technique undergoing its certification process, a rare access for any institution. Thanks to Kim Warren and BT, who were the first to welcome me into the space. I was so lucky to share so many days filled with labor, rat snuggles, and meals with the brave people who are actively working to deal with the aftermath of war and violence.

While in Cambodia, I benefited from a whole community of scholars at the Center for Khmer Studies, especially Krisna Uk, whose expertise working with military waste in the east of the country helped me with both practical and theoretical concerns. Thanks also to *Neak krau* Sisotha Ocur and Sreypich Tith, whose guidance through the language, books, and life at the center was vital to me. I also could not have written this book without friends in Cambodia like Cheryl Yin, Sarout An, and Frieda Kreth.

In addition to the labor of fieldwork, the writing itself was nurtured by a supportive group of scholars. It began while I was a graduate student at McGill, and I owe so much gratitude to both my fellow students there and the professors who guided my writing process and fieldwork process. Thanks to Lisa Stevenson, a brilliant mentor whose ideas and writing processes illuminated my concepts. She has always been an advocate and I will continue to look to her as a model for how to allow graduate students to navigate their own thoughts. Thanks also to Gabriella Coleman, a steady-handed influence on my writing and pathways during and beyond graduate school. Eduardo Kohn mentored me with many discussions about human and animal communication that helped me think through what it meant to write about such things. I also owe gratitude to the classes I took at the Social Science of Medicine arm of the Anthropology faculty. Tobias Rees's classes and discussions helped shaped the way I approach theory and teaching. Sandra Hyde's guidance in medical anthropology provided a good background for my work on this book.

My fellow students offered me their strength, creativity, and time, and most of them have continued to support the project as the book developed. It has been a pleasure to grow with them through the uncertainties of post-graduate life. I cannot thank my dear friends and colleagues enough: Doerte Bemme, for supplying me with magical thinking and treats; Julianne Yip for mochi trips and writing tips; Philippe Messier, for videography advice and care with theory; Adam Fleischmann, for company and readings on environments; and Alonso Gamarra, for the careful and sensitive thinking as I rush through my own with him.

As a postdoctoral fellow at the Institute for Genocide and Mass Atrocity Prevention at SUNY Binghamton, I was able to learn from Max Pensky and Nadia Rubaii. This book has been long enough in the making that the world has even lost some of its earliest readers, like Nadia. Nothing ages a project like the passing of one of its mothers, and I remain grateful to her early thoughts on it. The postdoctoral position allowed me to hold a book workshop over Zoom, due to the pandemic, and the feedback of the invited scholars was crucial to my work. In a sense, these were my first readers and gave real shape to its core, so many thanks to Sokhieng Au, Lucas Bessire, Faye Ginsburg, Radhika Govindrajan, Alexander Laban Hinton,

Ashley Thompson, and Zoë Wool. The topics I explore have also led me in various directions to those who study war, military waste, and their environmental effects. Without a doubt these are some of the most attentive and conscientious scholars, bearing witness to multiple devastations of violence. I have been lucky to overlap with them at conferences and symposia where they have commented on my work on this book, and while this is not a comprehensive list of this community, I am grateful that we, as humans, have people working on knowing and learning about atrocities and their afterlives. So, many thanks for the work and guidance of Eleana Kim, Kristina Lyons, Leah Zani, David Henig, Deborah Jones, Amahl Bishara, Bridget Guarasci, Ken MacLean, Josh Reno, Lisa Arensen, Diana Pardo Pedraza, Adriana Petryna, and Tam Ngo. This book has also been shaped by my time as a professor at the Department of Anthropology at the University of Oklahoma with the support of my research funds from there. So many thanks to my Okie colleagues, especially those who participated in the sociocultural anthropology writing group like Claire Nicholas, Elyse Ona Singer, Dan Mains, Kim Marshall, Camilo Sanz, and Lori Jervis.

In 2019, this book entered the Atelier Workshop series run by the brilliant editors Kevin O'Neill and Kate Marshall. I am so grateful to the Atelier series not only for their advice in navigating the publishing of the book but also for their talents when it comes to editing. The series introduced me to a community that has bolstered the book, including my original Atelier cohort and the close friends I have made over the series workshops. Thanks so much to a coven of inspiring and wonderful scholars like Sahana Ghosh, Namita Vijay Dharia, Sarah Besky, Tracie Canada, Laurie Denyer Willis, and Nomi Stone. To my dear hobbyists with a tarot deck of talents who have supported later drafts of its chapters, Emrah Yıldız, Kate Marincr, and Kaya Williams, I am so indebted to you not just for writing help but also for all the exchanges.

I am from a family of civil servants who have shaped me and the way that I approached this book and its research. My low risk adversity, perhaps a capacity that led me to minefields, comes from my father, Brian DeAngelo, who chose to fight fires because it was a steadier career than carpentry. My tendency to investigate trauma comes from my mother,

Gretchen DeAngelo, who was a rape crisis counselor and youth advocate for much of my life. My childhood was surrounded by four siblings who have inspired me with their own work as we have grown up together—Elysse, Abby, Toni, and Sam. This book was completed the year I became engaged to John Dao-Tran, who has supported it by reminding me that life is not just in the words we wrangle in our heads, but also in the ways we play.

Notes on the Language

Khmer words have been transliterated from the Khmer script to Roman alphabet. There are many competing systems for transliterated phonetics in Khmer and no standardized system, but the system used herein is based on the system in place by the Documentation Center for Cambodia, although I also add vowels when necessary for better pronunciation, as in the case of *sralanh*, which could be transliterated through DCC's system as "srlnh," and I use conventions in English such as *khmer*, which according to DCC's system should be "khmey." Other systems take into account more variation in diacritics (Antelme 2007; Thompson 2016), but for the ease of reading, I have used a simpler romanization. The Khmer words here are alphabetized according to this transliteration's letters, where *k* comes before *kh*. Unless otherwise noted in the text, all conversations are translated from the Khmer—first by me and if there were video and/or audio recordings, by a second translator who is a native speaker who reviewed the video and audio with me.

Transliteration	Khmer	Approximate Translation
anet	អាណិត	pity
aphika	ឧបេក្ខា	equanimity
bannha	បញ្ញា	wisdom
barami	បារមី	charisma, virtue, or in Anne Yvonne Guillou's work, potency or energy
barang	បារាំង	French, European, or white person
bon	បុណ្យ	merit
bong	បង	elder, older
chei	ជី	nuns
chol	ចូល	enter
chomteav	ជំទាវ	important lady
dauk	ដូច	equivalent or alike
kambab	កំបាប	sin
kap	កាប់	ax or cut
karuna	ករុណា	kindness or compassion
knea	គ្នា	each other
kraham	ក្រហម	red
krama	ក្រមា	scarf or towel
krau	គ្រូ	wise person, teacher
khnom	ខ្ញុំ	I, me
khmer	ខ្មែរ	Khmer or Cambodian
khvai	ខ្វៃ	kill
louk	លោក	mister
metta	មេត្តា	pity-love
menmen	មិនមែន	not at all, no longer
motita	មុទិត	loving joy or compassion
muk	មក	what
neak	អ្នក	you, or a person
prahmveher	ព្រហ្មវិហារ	brahmavihara (the four boundless of states of Buddhism)
praloeng	ព្រលឹង	soul or thoughts
satthi	សតិ	mindfulness
sangsra	សង្សារ	reincarnation, cycle of life and death
snaeh	ស្នេហ៍	love or attachment
som	សូម	please, beg

sorsai	សរសៃ	threads, string, veins, or filaments
sralanh	ស្រឡាញ់	love
thommachiet	ធម្មជាតិ	environment, nature, directly translated as "dharmic existence"
thviw	ធ្វើ	work, make, or do
tineh	ទីនេះ	here
trauw	ត្រូវ	need
utihir	ឧទាហរណ៍	for example
veay	វាយ	hit or beat
yeay	យាយ	grandmother
yon	យ័ន្ត	short Khmer word for *yantra*

1 New Choreographies

To find landmines with a rat, first you need a minefield.

The minefield must be prepped. Its flora must be razed. Sometimes this involves controlled burns. Trees and larger bushes get chewed to mulch by a specialized military brush cutter. The driver is protected by the vehicle's heavy armor. For the rest of the plants, deminers walk one by one in their heavy protective equipment, tearing them out. Then, once the minefield's ecosystem is dismantled, a charred field ready to be cultivated, the deminers break up the minefield into a grid of sections. The sections measure 200 meters squared. The contaminated rectangles must be surrounded by cleared corridors—this happens one rectangle at a time. You need cleared corridors because to find landmines with a rat, you need two people to guide the rat across the uncleared sections. The uncleared squares are bordered by measuring tape. The prepped minefield has become ready.

Second, you need a Giant Gambian Pouched Rat already trained to detect TNT, the explosive powder found in landmines. Third, you need two people who are trained rat handlers and deminers.

Each human deminer steps into a loop at the end of a rope—a rope that is strong and with a bit of elasticity, like the rope climbers use or a rope that might be used for an expensive dog leash. The two people stand across from each other, connected by the rope looped around their mirrored ankles or calves. The rat wears a harness with a latch connecting her body to the rope.[1] The two people hold a measuring tape loosely across the square, and the measuring tape is also connected to the latch. The rat walks across the square from one human to another. This way, when she stops because she has detected TNT, the humans can make note of the place on the contaminated section based on the tape measure: how many feet from the side of the square? How many feet from the end of the square? Where is the landmine? The humans map what the rat detects based on that gridded cartography.

This ecology—its flora, fauna, and even its human residents—has been reshaped by postwar operations of mine clearance. It is the result of aspirations toward disarmament, contaminated by military waste but reconfigured by human hopes for a postwar reality.

When a rat smells a waft of TNT through the soil, she stops and scratches twice. One of the humans presses a little training clicker, which the rat associates with a reward of banana. She enjoys hearing it. She is motivated to scratch the ground when she smells TNT because the association sends happiness and pleasure through her body. The other human makes a note on the grid map. The rat continues to the other side of the square where the second human awaits her. Then, the humans each take a step as one to guide the rat down the mine pit.

The choreography of landmine detection with rats seems complicated, but it takes only twenty minutes for a rat and two humans to complete scent detection for one pit. The rats do not die nor are they in danger of exploding—they are too light to set off landmines. The humans do not walk where the rats do, but their map allows a team to later detonate the minefield. This was developed recently and remains rare as a landmine-detection technique—first in parts of Africa like Mozambique, Angola, and Tanzania in the early aughts and then in Cambodia in 2015, where I witnessed it.

The new steps of landmine detection with rats have changed how some people move on a minefield in Cambodia. I had studied minefields there for half a decade before the rat entered the field. As an ethnographer, I relied on methods like qualitative interviewing techniques and in-depth, long-term fieldwork. I lived with people who lived and labored among landmines. Over the course of my studies, the people I worked with became friends and collaborators; I call them interlocutors to refer to this interlocution, these conversations that depend on their testimonies and generosity. In 2010, I conducted fieldwork in densely landmine-contaminated villages in the northwest of Cambodia. While there, my freedom to walk was curbed—I could safely move from home to home for interviews only via a heavy pickup truck that a friend skidded expertly through the wet mud of the rainy season. Being circumscribed by these limits made me interested in those who could walk across this dangerous land. For over fourteen months in 2015, I followed demining processes in general and then the introduction of new animal companions who helped human deminers. I was on the ground when the rats were first imported to Cambodia, and I witnessed how they changed movement in minefields.

During the implementation process of the new technique, the rats changed not only how people moved but also how people communicated with and related to each other in a fraught postwar ecology. The deminers come from histories of violence and military backgrounds. They had to learn to work together on a minefield with their fellow humans, who, in many cases, were their former enemies. The rats provided a way for them to do so by mediating their conflicts and suspicions.

I came to understand that I could not write about being human or nonhuman without understanding the warfare that gave rise to the forms of their relationships. The rats in the minefield showed me not only how war creates worlds through materialities, beings, and relationships, even long past its end date, but also how beings push back and reconfigure themselves beyond the limits of war's shape. This is the *post* in postwar ecology: a refusal of the circumscription of war worlding. The new choreography of landmine detection with rats allowed relationships to transform between humans and their human colleagues as well as between humans and the newly introduced rats. Such an introduction provided a kind of "first contact"—not in the colonial sense but a first exposure of these particular humans and these particular rats to one another. The rats described in this book learned not to tremble in the hands of a primate, while the humans learned to work with rats but also with their human enemies. These processes occurred in parallel, imbricating rats and humans who learned together but also depended on the violence that came before.

The minefields gave rise to these learning processes. They conditioned the possibilities of these new technologies and relations: war technologies—that is, bombs—gave rise to the invention of these rat-human teams. When I first conducted fieldwork in Cambodian minefields in 2010, I studied prostheses repair at a village clinic in the Sampov Lou district at the border of Thailand with the Trauma Care Foundation, a now-defunct Norwegian-Cambodian NGO. This northwest corridor was part of the K5 belt, the most densely contaminated landmine strip in the world, with over one thousand landmines per linear mile. Since my movement in the village was limited by the dangers underfoot, I spent much of my time with a former soldier, Hov, who had a Red Cross prosthesis.[2] We would sit in the small hut where he cultivated mushrooms behind a crop of garlic, and he and his wife told me stories about his time at war. He lost his limb in the battlefield during the 1970s. He never mentioned what army he fought for.

This silence is likely because he fought for the Khmer Rouge. The Khmer Rouge was a Maoist-Communist regime that took over Cambodia once the United States departed from Southeast Asia at the end of the Vietnam War. Cambodia was not unscathed by the Vietnam War and foreign interventions there. The United States dropped millions of bombs in the east of the country, despite the fact that Cambodia was ostensibly not part of the conflict between France, the United States, and Vietnam. Once the United States withdrew, it left a power vacuum, infuriating people. The aftermath of violence gave rise to the oppressive Khmer Rouge and its leader, Pol Pot. Pol Pot, called Brother Number One, ruled the Khmer Rouge with a paranoid governance characterized by forced work camps, forced marriages, torture, and death. The Angkar, or Organization, as it was called, separated families, recruited child soldiers, starved people, forced them into labor camps, and tortured and massacred millions. Pol Pot and the Khmer Rouge regime are responsible for a genocide, an estimated 1.7–2.5 million people murdered, at least a third of the country's population at the time.[3] That estimate represents a range since the country also underwent a famine at the time, leading to starvation and disease in addition to the murderous campaigns. The landscape has ever since been littered with unidentified bodies called "the killing fields," since the Khmer Rouge soldiers left their victims in mass graves.[4]

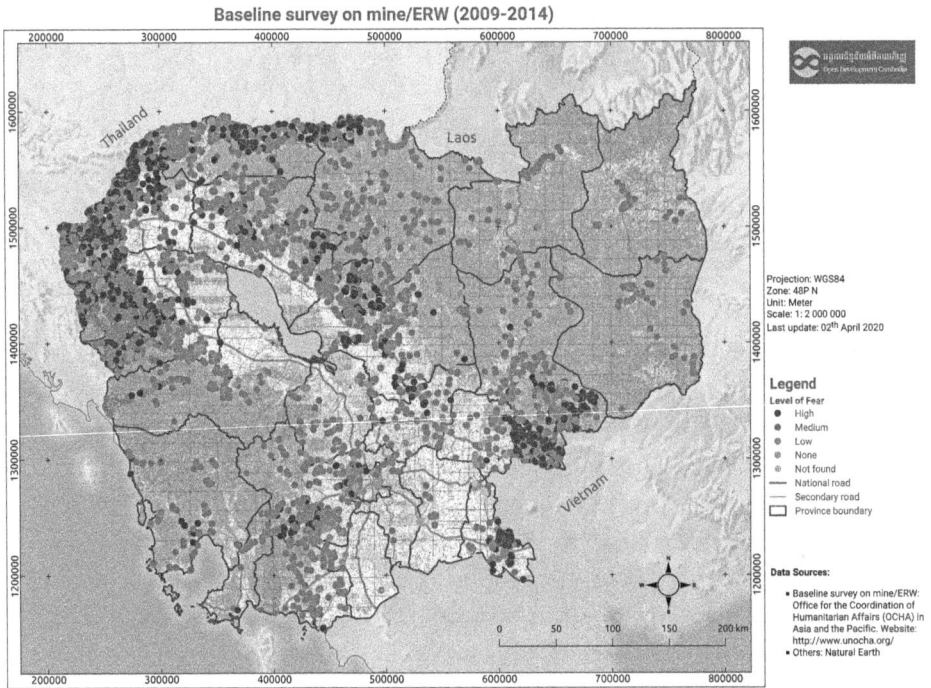

Baseline survey on mine/ERW (2009-2014)

Map of the Baseline Survey of Landmines and Explosive Remnants of War in Cambodia with levels reported by villagers and other sources (2020).

Today a majority of Cambodia's population is forty-three years old or younger because of the generations left dead. This means that most of the current population grew up after the Khmer Rouge regime ended in 1979 (Cambodia Demographic and Health Survey 2014). But this statistic fails to take into account the uneven distribution of that regime's end. Most of the deminers I know are from the periphery places, where fighting continued until the 1990s. Most of the landmines in Cambodia were planted from 1985 to 1989, when the Vietnamese-allied government installed a "bamboo curtain" against the invading Thai and Khmer Rouge along the Thai-Cambodia border in the northwest (Nhem 2014; Slocomb 2001). This area, called the K5 belt, remains the most densely landmine-contaminated region in the world, a 1,046-kilometer (650-mile) strip of land with up to 2,400 mines per linear kilometer (see map) (Landmine and Cluster Munitions Monitor 2016). After living and working near Hov, I followed those who walked across the minefields. I discovered that these minefield laborers had similar histories to Hov, who was a former combatant turned farmer.

The aftermath of civil war leaves communities full of blurred lines between innocence and criminality. The people employed to labor in the minefields, to solve the problems of these postwar ecologies, were produced by these wars. My interlocutors were the child soldiers who had been recruited as youths into the Khmer Rouge. Or they were those who fought against the Khmer Rouge. Or they had been both. Former Khmer Rouge soldiers like Hov told stories of being conscripted to the Vietnamese-allied army to fight at the border in the 1980s, after Pol Pot had been ousted. The soldiers who disarmed minefields were often the same soldiers who armed those mines in the first place. Like the minefield that presented a palimpsest of conflicts from the Khmer Rouge civil war in the 1970s to the Vietnamese takeover in the 1990s, a typical deminer's biography held multitudes, histories of being conscripted as a soldier to various sides of wars, shifting conscripts of Cambodia's long history of civil conflicts. The landmine itself was a material manifestation of those histories, a reference that disrupted the categories taken for granted in notions of stable relationships with each other. Landmines co-constituted being human in the postwar ecology because they materialized the actions of humans who had responded to each other during war.

It was only in 2015 that the rats entered this landmine-riddled field in Cambodia alongside a few people with troubled histories, but both the rats and the landmines had been around long before. The first rat for landmine detection was caught in a trap far from Cambodia in the 1990s as a response to another postwar problem. He must have trembled in the hands of the field technician. He would have been caught in a trap, drawn by a banana or avocado or peanut, lured from his familiar burrows atop which grass swayed and created nooks and crannies of shadows.

This rat must have trembled because he was out of his element, in the hands of humans, but he did not need to fear the landmines—not that he even knew what they were—because he was too light to set them off. The inside of that lab in Mozambique would have been more terrifying than a minefield for a Gambian Pouched Rat. Humans were the real danger to him. They were about to employ him as a postwar technology, a biological technology, as they called him and his eventual brood. He would be the first of his kind to be a technology for postwar decontamination efforts in the landmine-detection industry.

The first rat was caught at the behest of Bart Weetjens, a CEO, who exported him and his fellow rats to Antwerp, Belgium, where he introduced them to a lab environment. The rats were put in climate-controlled rooms where the lights blinded them and the humans smelled like warnings. This was the founding of APOPO, an NGO whose name was an acronym for the Belgian Anti-Persoonsmijnen Ontmijnende Product Ontwikkeling, which translates to Anti-Personnel Landmines Removal Product Development.

Weetjens, a self-described entrepreneur, Zen priest, and rat enthusiast, set out to train rats to detect landmines. Inspired by a research article about gerbils trained to detect TNT, Weetjens thought it might be possible to train rats to do the same. A rodent specialist, Dr. Ron Verhagen, had told Weetjens that the Gambian Pouched Rat, also known as the Giant African Rat or *Cricetomys gambianus,* was the rat best suited for his idea. The Gambian Pouched Rat, Verhagen said, was the best kind of rat for the job because of its size and its African origin. (This was long before APOPO's expansion to Southeast Asia.) The rationale was that, since it would be deployed in Mozambique and Tanzania, an African rat should be acclimated to those environments.

The rats were also large, the size of a small cat. The size would make it easier for humans to see when they indicated the scent of a landmine. Moreover, these were ground rats, not arboreal. Ground rats have a more discerning sense of smell than their tree-dwelling cousins.

But these rats couldn't do it alone. They needed humans (as well as bombs) to become landmine-detection technologies. So, they were trapped and imported to Belgian labs for testing and breeding. Unfortunately, getting the wild rats to reproduce in captivity proved difficult. The research team still needed to test the scent detection and trainability of rats in the timeline afforded them by grant funding but could not use their chosen wild rats. Undeterred, the scientists used an understudy, their old companion the Norway rat, for these first tests in the late 1990s.

While the Gambian Pouched Rats waited backstage, Norway rats, the common brown rats humans have known for centuries worldwide since the bubonic plague, learned how to smell TNT, associate it with a reward, and indicate to humans that they had smelled TNT to obtain the reward. Norway rats are already star performers in lab sciences (though also infamous across the world for invading grain houses and bringing pestilence to humans). Smaller than Gambian Pouched Rats, they were only part of these experiments so that APOPO could prove to funders that it was possible to train rats to detect landmines. Behind the curtain, humans kept trying to encourage the Gambian Pouched Rats to reproduce.

Meanwhile, back in Angola, the natural habitat of *Cricetomys gambianus*, the British princess, Diana of Wales, walked across a minefield. She was in the middle of a campaign to promote awareness of the global landmine problem, the aftermath of long civil wars waged decades before. Some of these wars were extensions of Cold War conflicts, heated civil battles partially funded by the Soviet Union and the United States, and others were postcolonial civil wars as empires fell or withdrew. The 1980s and the 1990s littered explosives across these battlefields while APOPO attempted to validate its concept of landmine-detection rat technologies. Landmines, in a word, were trending.

The most common antipersonnel landmine in Cambodia, the M14, is made mostly of plastic around a bit of metal that explodes when a human steps on it. You can hold it in your hand. It looks like a household smoke alarm. The M14 is also called the "toe popper," a morbid joke, since its "pop" results in an explosion that shatters a person's lower leg from their knee. The ingenious addition of plastic made it cheap, lightweight, and hard to detect. Developed by the United States, many M14 designs (usually unlicensed) today come from Myanmar, China, India, and Vietnam.

Long before plastic, the first landmines were metal spikes that soldiers would bury. Enemies would run across battlefields only to pierce their feet. Today, landmines explode when triggered by a footstep, but early landmines were functionally the same. The point is to maim the enemy, not kill them. The idea is that a landmine injures one soldier so that another soldier or even two must carry their fellow off the battlefield. Plastic made this process more likely. This is intentional in its design. They are also intentionally difficult to find. As she pushed the button of a dummy landmine to demonstrate the detonation process on her walk through the Gambian Pouched Rats' home pastures, Princess Diana commented on the intractable problem of landmines: "One down, 17 million to go." The switch to plastic in the 1990s made these landmines harder to detect. The change of material created a new problem and a new opportunity for innovation: TNT detection became more effective than metal detection. Deminers needed technologies that could detect the TNT itself rather than only metal—technologies such as rats.

Landmines' plastic ingenuity has made them a military technology that is both among the cheapest and, relative to its inexpensive primary investment, the most difficult to remove. Each landmine costs about US$3 to produce and $300 to remove. In terms of military weapons, plastic landmines provide the biggest bang for the buck. They are efficient and effective. Nation-states like the United Kingdom used to be very pro-landmine. As they became mass-produced, however, the soldiers on the ground, began to file complaints. Many landmine planters found them abhorrent, as Cambodian deminers who had formerly worked as planters told me.

Princess Diana advocated for landmine removal while geopolitical negotiations were underway to initiate a mine ban treaty. The Convention on the Prohibition of the Use, Stockpiling, Production and Transfer of Anti-Personnel Mines and on Their Destruction, known informally as the Ottawa Treaty, the Anti-Personnel Mine Ban Convention, or often simply the Mine Ban Treaty, was the result of six NGOs attempting to expand an earlier UN convention that restricted the use of indiscriminate weapons. Since 1977, a provision in the Geneva Convention has prohibited the targeting of civilians during war. The Mine Ban Treaty responded to the specific design of antipersonnel landmines, which had the unintended consequence of injuring civilians after war had ended.

In August 1997, the United Nations Mine Ban Treaty was signed by 121 countries in Ottawa, Canada. Landmines were banned. Not every country signed the treaty, but committees and funding streams opened up for the removal of these explosive remnants of war (ERWs).

Minimal metal mines—plastic war trash—required new technologies of detection that would pinpoint the bombs. Scent detection was already used with dogs. Bart Weetjens had seen an opportunity to patent a truly unique method for landmine detection at a moment of fortuitous timing.

If only the rats would comply.

.   .   .   .   .

A STEP-BY-STEP GUIDE TO LANDMINE-DETECTION WITH RATS

From an Anthropologist's Point of View

Step 1. Preliminary steps include rat socialization. After rats are habituated to humans, use clicker training reinforcement so that rats associate the smell of TNT with steps that lead them to a banana. Rat should learn to (a) smell TNT, (b) scratch soil twice, (c) hear click. The click becomes an equivalent to the taste of banana for the rat through intermittent positive reinforcement.

Step 2. Drive a military-grade brush cutter through a known minefield. The brush cutter has blades like giant weed cutters near its front wheels. It will tear down trees, cut down brush, and flatten small hills. It tills the earth so that the field looks like a freshly planted crop. The tilling is surface level so it does not set off the landmines.

Step 3. Detect landmines with a metal detector down a 200-meter strip. Detect until you can clear a space for pathways around a 200-square-meter minefield section where the rats will detect the TNT. Mark the border of this 200-square-meter section with two stakes and twine. Twine should mark the squares off so that the team can walk on safe paths in the minefield. The twine should also be strung with tape measures so that each meter is indicated on the sides of the minefield. (NB: To make the paths safe, place a flag wherever the metal detector gives an alert. Send in a detonation team to detonate the metal-detected explosives. Metal detection must be conducted so that there are two parallel safe paths where human deminers can walk sideways facing each other across 200 meters of contaminated land.)

Step 4. Remain in the safe, cleared paths. Put your rat in her harness.

Step 5. Let the rat sniff a small pouch of TNT powder and give her a banana morsel when she scratches the earth twice. This is a pre-test, letting you know if the rat is ready to detect landmines in the pit. It also serves to reinforce the rat's association of TNT powder, scratching, and banana.

Step 6. In the cleared paths around the square pit, two deminers stand across from each other. (They should set up their human-rat-human rope configuration in the border outside the mine pit.)

Step 7. Each deminer puts one leg into a looped rope that connects to the harness. One deminer loops the left leg and the other deminer loops the right leg so that they are mirror images of each other. They also string a tape measure through the harness loop so that they can mark how many

Figure 1. The typical setup of a landmine detection rat team, with two deminers standing across from each other with the rat attached to a rope that is attached to an ankle or leg of each of the humans. Here are two deminers on the practice minefield near Siem Reap. Screenshot from film by author, 2015.

feet into the pit the rat has scratched, indicating that there is a bomb. The deminers will later create a map of the pit so that the number of meters both into the pit and along the pit will be noted when the rat scratches twice. (See figure 1.)

Step 8. Allow the rat to walk across the pit to the other deminer.

Step 9. Allow the rat to walk back across to the first deminer.

Step 10. On the rat's return trip, the pair of deminers will mirror each other in one sideways step to the left or the right, walking down the mine-field path.

Step 11. Repeat steps 8–10 until the entire mine pit has been mapped. Hand the mine-pit map over to detonation teams to safely detonate the landmines the rats have detected.

These steps are new for landmine detection and they produced "easy" or "difficult" rats. The *Cricetomys gambianus* who are difficult don't fulfill the expectations of the humans seeking to transform them. The first dif-ficult ones were the ones who did not reproduce. Eventually, the scientists found easier ones who mated in captivity, became pregnant, and bred a litter who would become easy rats. Later on, difficult rats were defined as the ones who were slow in the minefield or were biters. The new rats being born learned how to restrain their instinct to tremble in human hands through acclimation from an early age. Being a technology rat means being held and stroked by humans as a pup. Wildness persisted in some of these rats despite their domestication. These difficult ones did not like being held and bit the fingers of humans they found intolerable.

Most of these rats do not tremble when you hold them. They are born in a litter of one to five pups. They squirm and, even though adult *Cricetomys gambianus* are mostly blind, the pups cannot even open their eyes. Humans stroke and hold the rats to habituate them to the smell of large furless mammals. After a few months, scent training begins. The first *Cricetomys gambianus* were trained on the landmine-detection technique in the early aughts, long after the Mines Ban Treaty was complete, long after Norway rats had proven APOPO's landmine-detection method, but during the global trend toward landmine decontamination efforts.

While *Cricetomys gambianus* was being trained to become a new companion species to detect landmines in places in Africa like Angola, Mozambique, and Tanzania, most contaminated places in the world relied on other methods for ERW detection that involved using metal detectors. However, to detect minimal metal mines, the plastic war trash, humans needed a detector that could discern the landmine even when it was plastic. Thus, landmine-detection dogs who could sniff out TNT were first used for this purpose. The rat, just on the horizon, was also perfectly suited for detecting plastic war trash.

Finally, in 2015, in Cambodia, after Princess Diana's walks, after the Mine Ban Treaty, after prepping the rats, a candidate sat on one of the heavy wooden chairs set before a heavy wooden desk. She answered a series of questions about her landmine-detection experience, which was often based on the candidate's past as a soldier or, in Moch's case, on her experience living in a minefield and having a father who was a military police officer. Growing up in the minefields near Preah Vihear, she said, had made her familiar with landmine safety and had motivated her to apply for a landmine-detection job. Her father had recommended her for the job.

The most important test for the candidates was not their experience working with landmines nor their connections nor their passion for decontaminating military waste. Behind Moch's chair, a door swung open and a deminer entered with a plastic transport cage holding a massive three-pound rat. This test assessed how she touched the rat.

Moch cringed only a little.

Her potential employers, all hired by the NGO APOPO to implement the new biological technology, sat across from her. A British woman, a man from New Zealand, two Tanzanian consultants, and two Cambodian men, one of whom acted as translator for the foreigners, watched Moch's response intently. She reached out for the rat.

"This is Simon; he is our most proper rat," one of the Cambodian men, Hien, said. He slid the grate from the shallow container and passed the rat to Moch's open hands. She took the rat a bit clumsily, cradling him in her arms as if he were a human baby. One of the Tanzanian men modeled the way she should hold her palms out to allow the rat to have a stable surface as he explored his new surroundings with trembling whiskers and hesitant steps.

Her hands spread wide and the rat swept its head from side to side. The rat was soft in her hands and she petted his back. Simon's claws held tightly to her skin.

"Moch was good," Hien told me later, "she was very gentle with Simon, and he is a gentle rat."

Both Moch and Simon built up good reputations in the landmine-detection organization where they worked after that first encounter. They both were considered "proper" and "hard working" when it came to landmine detection. They were featured in multiple promotional pamphlets and videos as APOPO, the Belgian NGO that implemented the patented landmine detection rats in Cambodia, in partnership with the governmental military agency Cambodian Mine Action Center, sought to publicize the new technology, the landmine-detection rat.

This first contact between Moch and Simon was a move not only toward new technologies and new techniques but also toward new relations. Moch, along with nine other Cambodian deminers, had been hired to be trained to use seventeen newly imported rats as landmine detectors for the first time in 2015. Most of them told me that they had never liked rats before they met the APOPO rats. These rats, specifically labeled biological technologies by the two organizations in charge of their implementation, required the use of techniques that completely subverted the ordinary movements of deminers on a minefield and, in so doing, disrupted the minefield's "ordinary affect" (Stewart 2007) as well. In highlighting these disruptions, I not only want to document and explore the transformations of relations between beings on the minefield, but to examine the minefield as a *postwar ecology*. By thinking through war as integral to the relations and transformations taking place, I situate the beings on the minefield (human, animal, and even spiritual) as both products of postwar relations and simultaneously transformative of such militarized landscapes. The rats push back against the materialities (and subsequent ethos) of militarized landscapes just as the militarized materialities of bombs created them in the first place. This (re)configuration occurred for the humans as well. Postwar itself is an aspirational concept that must be integrated into understanding these (re)configured relations. Thus, perhaps surprisingly, moments of trust, love, and affection were key to my understanding postwar ecologies.

Understanding postwar ecologies in Cambodia meant contextualizing them with the concepts and practices of diffuse personhood that comes partly from Buddhist philosophy and partly from the need to mediate after shared war traumas. That is, a person's self is not manifest in individual will and individual feelings about her surroundings when it comes to Buddhist theories of mind. Instead, selfhood is produced through multiple ways of relating to humans and nonhumans such as spirits (Eberhardt 2006; Cassaniti 2015). This kind of diffuse personhood blurred definitive agency of violence in war stories I heard, too, as if to hold everyone accountable for the atrocities of the past. To put these surprising moments of love and affection in context, I drew from how most of my interlocutors portrayed senses of diffuse personhood that extended relationality to rats, spirits, and their former enemies. This diffusion troubled ideas of justice but allowed former enemy combatants to work with each other, to transform each other. The rats helped these processes along.

In considering these relations, one cannot overlook that, in the context of demining, rats are both technologies and beings. The claim that rats are a "technology" comes from the demining industry itself, where rats and other creatures like dogs, bees, and fungi are called "biological technologies" for their use in detection techniques (Habib 2007). In a review of literature on materialism and its relation to ecological anthropology, Tim Ingold (2012) observes that humans and other beings are *materialities and technologies*. He suggests that separating material and ecological frameworks risks ignoring how materialities (like explosives) are constitutive of the relations that give beings form. In parallel, Kim TallBear (2011) notes that the very ideas of abiotic/biotic, human/nonhuman, and "interspecies" have taken for granted the givenness of categorical distinctions they impose on relations. Drawing from the ways in which Cambodian deminers understood rats as both technologies and beings "confound[s] the Western animacy hierarchy" that disrupts taken-for-granted perceptions of a minefield (TallBear 2017, 180).

For instance, it may be tempting to suggest that the rats were *only* to be understood as beings, since their handlers never described them solely as "technologies" or "machines" (as rat handlers seem to have done elsewhere [Lee 2021]). As Nomi Stone (2022) has pointed out, it is quite common in US military settings to understand people as kinds of technologies who sometimes push back against their functional expectations. Technology in its precise definition can be conceptualized as something that is purely functional with no agency, despite how human practice may complicate this. Although the rats served a functional purpose due to their unique evolutionary "design"—that is, their biologies—just as humans did in a minefield, they, like the humans, also had agency that had to be engaged with to ensure that they served their functional purpose. When the humans engaged with that agency, they felt real delight and pleasure in their rat companions who, in turn, seemed to appreciate the cuddles, petting, and good snacks. That kind of agency—desires beyond their functionalities—afforded a new affect on the minefield, which was reflected and foundational to their functionalities, the technological categorization, in the first place.

One of the first times I met the rats was in a kennel with the Cambodian vet Tokla as he was being trained. Tokla had already spent the night with them in the kennels, eschewing an evening with the other humans to sleep on the floor while the rats engaged in their nightly playtime. He had found it difficult to sleep because the nocturnal rats had been chewing and

Figure 2. Frederic stands next to his chew toy, Siem Reap. Screenshot from film by author, 2015.

nesting all night, responding to the fact that they had just been moved to new crates with new smells. The kennel held a total of sixteen crates; each crate housed a rat.

Frederic was the only one who stayed awake in the late afternoon (figure 2). Together, we watched Frederic gather straw frantically to pad his crate, perhaps trying to line it the way he might insulate an underground burrow. Spit-matted straw fell through the wires.

"Which one is your favorite?" asked Tokla.

"Frederic," I said, sympathizing with his Sisyphean task.

"Good choice," Tokla smiled.

As favorites, beloveds, and friends, the rats, themselves (re)configured by war, conditioned new possible relationalities for human colleagues who were also former enemies. The ways in which the rats altered the movements of landmine detection brought deminers into new practices of coordination. No longer singular individuals, they walked in pairs with a rat between them. Donna Haraway uses the concept of "string figure" to draw attention to the ways in which worlds are (re)configured through processes that entangle and disentangle actors, categories, and practices.

For Haraway (1994, 63–65), string figures are "imploded" materialities and metaphors, interlinked so that each configures, refigures, and reconfigures the other. I like to play with this idea of string to describe the collapse of the individualities on the minefield as well as the blurring of boundaries between human and animal, technology and kin, and enemies. The minefield's new choreographies, accompanied by string and bio, have also changed the ways humans see each other. Instead of individuals walking one by one, rats bring pairs of deminers with histories of enmity to share each other's attention to a nonhuman companion, face each other, and learn to move together.

The rat allows these emergent transformations to take place. In human-animal studies, scholars have pointed out that the interactions between humans and nonhumans produce new modes of being—that is, new practices of what it means to conceptualize "human" and "animal." Haraway affiliates herself with the philosopher Vinciane Despret, who describes this process of co-constitution as "an 'anthropo-zoo-genetic practice,' a practice that constructs animal and human" (2004, 122). Despret describes how experiments in the 1960s portrayed unseen communication between lab rats and human experimenters. When a human experimenter expected that a rat was from a "bright" group, the rat solved the maze puzzles better than the rats labeled as part of a "dull" group. She reframes these experiments not only as human labels and unconscious expectations that alter a rat's performance but also as ways in which the rat and the human trust each other and ways that the rat responds to the human's expectations. Despret links trust to love and suggests that trust is a matter of expectations that are met within relations for both animal and human (114). Trust itself, though, was a novel expectation between beings on a minefield before rats entered it.

In this way, co-constitutions are about becoming responsive to expectations. Such trust is a new way of "becoming" a rat—what Despret calls "being-with-a-human" (122). According to the design of landmine detection with rats, "being-with-a-human" was being-with-two-humans or perhaps being-with-a-human-with-a-human or even being-with-a human-with-a-human-with-a-rat. This becoming is not separate from the relations around it. This embedded becoming helps me outline what I mean by *ecology* when it comes to the postwar ecology of the Cambodian minefield. As scholars in environmental relations have pointed out, an *ecology* serves as a useful term to undermine the divisions between human and nonhuman. The anthropologist Alex Nading (2014) has pointed out that to separate nonhumans from human practices that shape them, such as the proliferation of mosquitos and mosquito-borne illnesses as influenced by waste management and public policy, results in an incomplete story. Likewise, Radhika Govindrajan's (2018) study of nonhuman animals and their intimacies with humans must also be grounded in the specifics of the place and time where the nonhuman animals as well as their biological attributes are situated. Rat-human relations and human-human relations in Cambodia were embedded in the ecology of a landscape riddled with explosive remnants of war. Thus, if war constitutes people, as Zoë Wool (2015) posits when she depicts how war itself circumscribes identity and healing, war also constitutes landmine-detection rats. But they, in turn, reconstitute the minefield relations via their entry as a unique actor in the minefield and in the demining industry as a whole (DeAngelo 2018).

Frederic insisted on building a nest when he couldn't. Veronica hated her harness. Merry squeaked in his sleep, prompting one deminer to infer that he had nightmares. Similarly, Simon became another deminer's "little sister," despite being the largest rat of the platoon—and a male. When I asked Moch why she called Simon "little sister," she explained: "Love. I know he's an animal and that he is male, but I just feel like he is my little sister." By giving careful attention to the rats and by telling stories that allowed them to share in each other's attention, the deminers learned to relate to each other differently, too.

·   ·   ·   ·   ·

Let me describe a typical minefield without rats working toward the goal of decontamination. Depicting a typical minefield without rats will allow me to more clearly contrast the rat detection technique with traditional landmine-detection methods.

Even under the dawn's light, the air felt muggy on the minefield. The stocky mountain of Bunong overlooked pale tapioca stalks, toppled over by vehicles that had flattened the ground in preparation for the detection platoon. Rose-colored clouds bathed the mountain in mists that stretched over the rice fields. In the eastern distance, I could see the gilt turrets of a pagoda and from that direction I could hear the chanting of monks who greeted the day.

I commented on how beautiful it was.

"Yes," the supervisor agreed, "minefields are always beautiful. When you want to find a landmine, you look especially carefully under trees or by rivers. That's because an enemy will rest there. When an enemy is off their guard, they will sit and relax or try to get a drink of water. Then, the landmine will explode while they are resting."

The platoon leader and I looked at the row of deminers. Each person stood hunched, arms held in tension as they swept a metal detector slowly before them. After each sweep, they took one step. They stood about three feet apart. Each step seemed laborious. They all wore helmets and masks and aprons of bulletproof materials. Their walking was slow. The sun reflected off the long plastic visors. No one spoke. It felt as though we were all waiting for the silence to be disrupted by the ominous beep of a metal detector. Or worse.

The supervisor still used the word *enemy* even though the wars that had left explosives in the ground had long ended and the contemporary victims of these explosives are usually civilians—villagers. The sensorial encounters of landmines—specifically, their invisibilities—perpetuate categories, persons, of war. They are hidden and they parallel hidden intentions of people, especially coworkers on the minefield, and they each must be hypervigilant when it comes to discerning them.

Cambodian minefields hide potential violence beneath bucolic beauty. Landmines are buried beneath crop soil, beneath vibrant vegetation, beneath trees and near ponds. As the supervisor told me, soldiers who buried landmines often chose a relaxing spot to hide the explosives. He knew this because he himself had planted landmines as a soldier. As a soldier, he knew his enemies would need a shady spot to rest or fresh water to drink. Because of this, landmines were in places that inspire peacefulness. Peacefulness was not a feeling to be trusted. Any tranquil relations came under suspicion not merely because of the explosives underground but also because of the legacy of war relations in demining platoons.

In contrast, trust between a deminer and a rat was necessary. For example, in their first encounter, Moch had to trust that the rat was not a pest, that Simon the rat would not nip her fingers. And, as a prey animal, Simon the rat had to trust that Moch the human would handle him gently. This trust, I came to find out, altered not only the relations between humans and rats but also the relations of humans with humans and humans with spirits. A learning and transformative process became possible through the postwar ecological relations themselves (betwixt and between human deminers, spirits, and rats).

Before that transformation, minefields had been places that reinforced mistrust. Aspirations toward attaining postwar in them had been limited because they were populated by actors who reinforced war relations and enduring enmities. On a minefield, I was told, you cannot trust anything. Indeed, in these postwar ecologies, I found shadow stories, lying clouds, and murderous spirits. The rats became surprising sources of love and trust in a landscape otherwise rife with suspicion. This book will draw you through the process of how deminers learned to trust landmine-detection rats and, in doing so, moved toward the aspirations of postwar. I begin with shadow stories and mistrust and then introduce the rats and the rat handlers who become responsive to each other, after murder and through love.

2 Shadow Stories

Over the course of my fieldwork, in addition to spending most of my time with the platoon who trained on the original landmine-detection rat team in Cambodia, I toured over thirty minefields. They are characterized by silence. And yet, like the visual tranquility of a rice field with hidden bombs, this silence represented a paradox: silence indicated safety but was also laden with the anticipation of violence. This paradox, or, as Eleana Kim (2022) says to trouble the very idea of a paradox, this different sense of peace also reflected how people formed friendships and spoke about themselves and their experiences of war. When you could not trust the ground you walked on, you could also not trust the past, which became blurry, with the landmines as reverberations of war in the contemporary that had ended years earlier. I saw this in who owned (or did not own) their actions of war—that is, I saw it most clearly when agency and accountabilities became muddled. The ways in which Cambodians spoke to me of war diminished or redistributed agency and accountability for past actions that they had both committed and suffered. This evoked a general feeling of mistrust on the minefield but also offered a way to navigate having to work with former enemy combatants.

My entry point into this field must be troubled by acknowledging the colonial role of anthropology generally and in Southeast Asia more specifically. The tendency of anthropologists to Orientalize and other has confined people into categories of savagery and exoticism, but violence is never exclusively local (Moten 2003). The former combatants I spoke with had committed violence in the past because of wars carried out in a global context of Western supremacist conflict that interfered in regional politics in a postcolonial attempt to maintain world domination. Indeed, the remnants of war in Cambodia are the product of US-based engineering innovation and many bomblets were dropped by the United States onto Cambodia—the fields in the east of the country hold millions of them thanks to the United States dropping explosives there during the Vietnam War. As a white cisgendered woman, my appearance and gender placed me as part of that legacy, and as an American citizen, I had a privileged status that the Cambodians I knew were aware of and commented on. They would describe the United States as a "strong country," and I know that they would have responded to me differently, most likely more favorably, than to another researcher from another part of the world and a different race.

I also want to acknowledge other power dynamics present in working with my interlocutors. While my gender, nationality, and race allowed me access to situations through power, being gendered as a woman who appeared young also sometimes meant I was sexually harassed and dismissed, especially in the militarized environment of a platoon of deminers and in interactions with former combatants. Sometimes this dismissal worked to my advantage because interlocutors felt comfortable telling me secrets they would not tell more powerful people. Sometimes, the sexual harassment kept me from participating in certain situations since I endeavored to keep myself safe. For instance, while I spent my days with the deminers, I did not stay overnight with the platoon in the minefield or the rat kennels. I followed the daily activities of deminers, occasionally attending social events such as parties in the evenings when I had a clear way of getting home.

My schedule followed the platoon's schedule: I attended their training with the rats in the mornings, I had meals with them, and I attended their courses in the afternoon. After receiving some requests, I also began teaching English to them in the afternoon. The rats would begin a few hours of work in the morning around dawn, about 4:30 a.m. At 4:00 a.m., I would meet the platoon supervisors and the Tanzanian consultants who had been hired to train the Cambodian platoons at Demining Unit 6 (DU6), which was the office of the Cambodian Mine Action Centre (CMAC) in Siem Reap. Then we drove 30–45 minutes to the kennels. Around 10:00 or 11:00 a.m., depending on the heat, the deminers would stop the practical lessons and have lunch. In the afternoons, they would have "theory" lessons, which were lessons on the various types of bombs, detection equipment, rat behavior, and safety protocols in the minefields.

The platoon members were hired in different phases. At first, even before the rats arrived, the Cambodian platoon supervisors, Hien and Chann, established the safety protocols and set up the minefield with Liz, the British NGO manager who had left her job at Handicap International to oversee the rat-technique implementation. Deactivated landmines were to be buried in the gridded structure of a minefield where the landmine-detection rat technique would be deployed. Next, the Tanzanian consultants, Mohammed and Daudi, arrived. Then, the rats arrived. Once the rats arrived, both the NGO and the government hired their rat handlers through the interview process. Five months passed before the field was up and running; I was present during this time as the training field and its protocols were prepped.

As a visual anthropologist, I conducted participant observation with a video camera, which allowed me to capture animal behaviors more closely and to provide a mode better suited to collaboration on ideas with deminers who do not read or speak English well. We have been sharing ideas ever since my filming in 2015, and as of this writing we are collaborating on the story of landmine-detection rat training for a short feature film. My film and my writing work together iteratively, and both attend to the sensorial encounters of what it was like to learn to love rats in a Cambodian minefield.

I also want to note that I have often been confronted by academic colleagues who assume that my interlocutors are at a distinct disadvantage in their own communities. Although deminers are stigmatized because of their association with death, violence, and the government, they hold positions of power, too, due to their participation in the Cambodian government and its military. These deminers are highly experienced and highly skilled workers.

For example, once at a conference, I showed a short video of deminers doing their work with landmine-detection rats. "How did you reconcile," someone on my panel asked, "going to the community and holding such an expensive piece of equipment that contrasts with the low-tech manual labor of landmine detection?" The question surprised me because it made me realize I was not being clear about the fact that landmine detection is a highly skilled form of labor. And, in fact, the rats themselves cost thousands dollars more than my video camera, which at the time cost US$4,500. A landmine-detection rat cost US$8,000 per year to feed and engage in continuous training. The deminers who worked with the rats were relatively elite in their field, and they also held authority over villagers, both because they were likely to have had combat experience in recent wars and because of their connections to the government. At the same time, their past shadowed their present.

One of the first former combatants I interviewed had joined a Buddhist monastery. Buddhism and language offered ways to express a kind of muddled agency when it came to the acts committed during war. When I asked him why he had decided to become a monk, the Venerable[1] usually answered me without speaking about his desires or his spiritual vocation. This made sense because to be a good Buddhist, you attempt to do away with your individual needs and wants, but he did say, "During the Khmer Rouge, there were no monks. And so, I was forced to go to the work camps. Then they later let me move to the factories where I made clothes for the Khmer Rouge soldiers. After the war ended, there were a few years where people could become monks. . . . My family encouraged me to join the monastery to make up for what [I] did."

The Venerable seemed to explain that his actions during the Khmer Rouge had caused him to dedicate his life to monastic practice. But this explanation remained unclear. When he said, "Som bon chol muk thviw tineh," I have translated it as "to make up for what [I] did."[2] I inserted the pronoun *I* in brackets because Khmer, as a pronoun-drop language, presents its subjects in an ambiguous form. In Khmer conversation, the speaker does not normally clarify objects or subjects, making context especially important when communicating. Khmer is considered a "radical pro-drop" language, a rare trait among languages but common throughout Southeast Asia (Haiman 1999).

"Pro-drop" languages often have verbal agreement patterns to clarify their pronoun subjects, as in Spanish and Italian, where a subject is distinguished as first, second, or third person via its verb conjugation. By contrast, Khmer is a language with radical pro-drop tendencies—that is, it has no verb agreement to indicate who the subject of a sentence is (Simpson 2005; Bisang 2014). This, I was told by native Khmer speakers, leads to confusion in conversations, but also to jokes. For example, the name of a fish curry noodle dish means "hand-fed noodles." Because of the radical pro-drop, the name can indicate either that the eater feeds herself or "you feed me noodles." Because no one ever feeds noodles to anyone unless they are small children, people joke that the noodle dish is actually called "you lie noodles," because in Khmer, the verb *to lie* rhymes with the verb *to feed*. The joke usually ends with "But I'm not a liar!" and someone holding someone else's noodles with a feigned intention to feed the person eating them. Ambiguous actors in stories and jokes like this paralleled ambiguity over whom one could trust on the minefield.

Perhaps a better translation for the Venerable's assertion would be "to make up for what was done."

The Venerable's discussion of "what was done," however, omitted direct reference to violence and who did it. His participation in the war through making clothes for Khmer soldiers is the only explicit part of the story. This, to me, sounds much less violent than other horrors that occurred during the genocidal regime, which raises questions of what *was* done and who did it. It seems significant that the Venerable describes his reason for going into the monastery as a choice made by "a nexus of kin relations" rather than an autonomous, individual decision (Strathern 1987, 272). He could very easily be making merit to pay off karmic debt not only for himself but also for his family members, who were perhaps more involved with the Khmer Rouge cadre's violence. In Buddhist practice in Cambodia, it is common to send younger men in the family to the monasteries to build *bon*—that is, merit—to make up for the whole family's bad karma (Zucker 2013, 171). This takes relationality to its logical conclusion: one person's good act is everyone's good act; one person's violence is everyone's violence.

The Khmer Rouge was survived not only by victims but also by people implicated in the violence of the regime's atrocities. The motivation behind the Venerable's choice to pursue monastery life was hidden by the unspecified nature of karmic debt. It is possible that the Venerable was paying off his family's debts rather than his own. But the true actors behind what was done remained unspecified throughout our conversations. This unspecified agency behind unspecified actions that someone needed to make up for may indicate that the Venerable understands his and his family's actions as intimately connected, even shared. Exactly who engaged in violence is a shadow story, an unspoken story behind the story shared aloud. It calls into question what it means to be the "real" actor behind an action. Perhaps it does not matter if you committed a crime when your brother's actions (criminal or not) have karmic consequences that implicate you.

Uncertainty about the nature and the agents of violence, like the Venerable's uncertain actors and actions in his story, was not unusual during conversations with deminers in the minefields. In this chapter's stories, I describe who my human interlocutors were and how a certain kind of distributed agency was key to their self-expression. Who are the people who entered the postwar ecology of a minefield as deminers and how did they describe their agency and actions of the past? By describing these in a way that conveyed uncertainty and untrustworthiness, they also allowed for a kind of distributed accountability that would help navigate untrustworthiness of their coworkers. While such uncertainties might consequently implicate everyone in terms of accountability for historical and personal violence, they did not eliminate suspicions of each other among coworkers in the minefield. Indeed, they may have even exacerbated such suspicions, breeding an atmosphere of untrustworthiness. Deminers would whisper about espionage by the government and among colleagues. They understood their affiliations with either the NGO or the government as making them rivals of a sort, working together in a forced institutional alliance. Their interactions with each other were often laced with unspoken fears.

I call these silences shadows behind stories to play with the sensorial encounter of something that obscures sight but also reveals darker shapes. You could not trust what you heard and saw, but you also could not trust people to listen to your stories in ways that would not be used against you.

War shaped not only the human actors and their relationships on the minefield but also the very ideas of agency. Untrustworthy human actors and relationships were inextricable from the postwar ecology of a minefield.

.　　.　　.　　.　　.　　.

Princess Diana's campaign against landmines depended on the presumption of the innocence of landmine victims. Innocence here takes an ethico-moral framework for prioritizing aid and asylum to those in the developing world from a Euro-American context (Ticktin 2017, 2016). By discussing landmine victims in terms of their worthiness, some people inevitably become categorized by implication as "unworthy" of aid, thereby justifying a lack of action. In images from the media about Princess Diana's activism, amputees gaze at her with adoration and need. They are often thin, on crutches, and pictured just below the perspective of the camera, so as to look up with yearning at the viewer. Or they are babies in a clinic who, according to her narration, are "incredibly well-behaved." Depicting landmine victims as innocent bystanders, like the domestication of *Cricetomys gambianus,* is a kind of transformation. Decontamination efforts, as they were called, often occurred amid war. In Cambodia, new minefields were still being installed until the 1990s even as the first landmine-detection platoons were being trained.

While children and pacifist villagers can be victims of landmines, many of the people affected by landmines are villagers who practiced war in the former battlefields. When advocates wrote on their behalf, the experiences of most villagers in those battlefields were glossed over to emphasize their current lives in a time of so-called peace. The landmine itself acted as a material manifestation of a soldier's war practice, a distribution of agency in material form (Gell 1998). Its materiality disrupts the categories taken for granted in ideas of stable relationships with each other. Thus, landmines co-constitute being human because they refer to other beings who have responded to each other, such as the distributed agency of soldiers who planted them before.

One deminer, Aki Ra, the cofounder of a nongovernmental organization (NGO) called Self-Help Demining, explained his complicated military

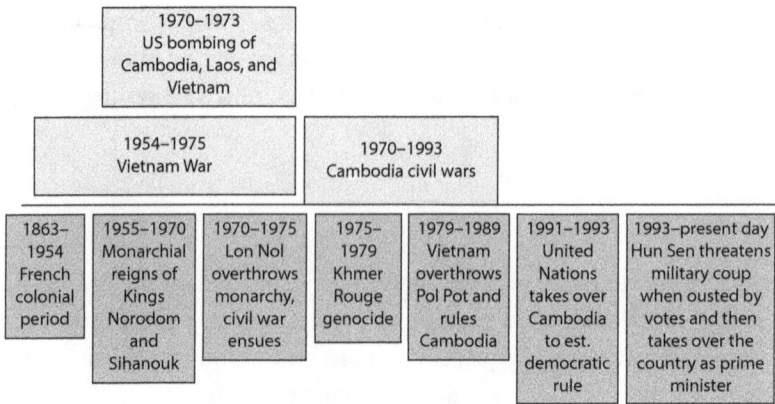

| 1970–1973 US bombing of Cambodia, Laos, and Vietnam | | | |
| 1954–1975 Vietnam War | | 1970–1993 Cambodia civil wars | |

1863–1954 French colonial period	1955–1970 Monarchial reigns of Kings Norodom and Sihanouk	1970–1975 Lon Nol overthrows monarchy, civil war ensues	1975–1979 Khmer Rouge genocide	1979–1989 Vietnam overthrows Pol Pot and rules Cambodia	1991–1993 United Nations takes over Cambodia to est. democratic rule	1993–present day Hun Sen threatens military coup when ousted by votes and then takes over the country as prime minister

Figure 3. Timeline of regime changes. Timeline by author.

past in a more open way than most of the deminers I encountered. His history of conscriptions follows the regime shifts of the twentieth and twenty-first centuries in Cambodia (see figure 3 for a simplified timeline). He described being a child soldier during the Khmer Rouge, being conscripted in 1979 to fight against the Khmer Rouge, and then being trained as a deminer by the United Nations in the postwar period. He once visited the landmine-detection rat platoon as a potential business partner. As he exited the building that overlooked the mock minefield, set up with twine boundaries and stakes to indicate where the "cleared" areas were, a few of my friends from the landmine-detection rat training program walked off the training ground to meet him. When I asked them how they knew him, they said, "We used to work together before demining."

This work "before demining," was combat experience. This meant that some of the deminers had been soldiers with Aki Ra, but with what army? All of Aki Ra's previous work was military. His long history in landmine detection comes from his long professional experience as a soldier. Aki Ra narrates his autobiography in a way that preempts any doubts about his ability to teach farmers how to demine, while also speaking to the complicated conflicts that have divided people.

On his website and in the Landmine Museum, a tourist attraction near the tenth-century temple site of Angkor Wat, Aki Ra describes being a soldier in all the armies of Cambodia's recent wars: a Khmer Rouge soldier who was taught to plant landmines as a boy; a youth who was forcibly conscripted by the Vietnamese to fight against the Khmer Rouge; and in 1991, a soldier for the newly Cambodian-run government invited by the UN to learn how to detect and detonate landmines. Three deminers told me that the UN had also recruited them in much the same way. In part because of its soldiers' prior experience with planting landmines, the Cambodian army was a common place of recruitment for the disarmament efforts of the United Nations Transitional Authority in Cambodia. All the deminers knew Aki Ra's name and smiled when they saw him. Like him, most of the deminers in the rat-training field had transitioned through various armies. Some were former Khmer Rouge soldiers; some had fought for the Vietnamese during the 1980s; and some were from the Cambodian army. These armies eventually reformulated into one large national military: the Cambodian army, composed of soldiers with splintered loyalties. In the national army it was thus common to find a former comrade-in-arms from a different army, but it was also common to find someone who had belonged to an opposing army.

When Aki Ra came to see the technological innovation of the landmine-detection rat, he had already been clearing mines without such technological innovations and without professional training from the UN. Aki Ra and other villagers claimed that they began removing landmines even during the civil war of the 1970s and under the Khmer Rouge to clear land for their personal use. Although villagers continue this kind of informal demining, these practices were later professionalized when the United Nations took over. Starting in the 1990s, professional demining activities in Cambodia were conducted using simple metal detectors and military expertise, including the expertise of those who had planted the mines (Bottomley 2003). This career trajectory of soldier to deminer was frequently portrayed to me by deminers. More generally, in NGOs and government demining organizations, the narrative of such a trajectory depicts professionalized expertise. For example, Aki Ra's demining organization, Cambodia Self-Help Demining, focuses on teaching villagers who live in minefields to detect and detonate landmines. Since the deminers in Cambodia Self-Help Demining are usually career farmers, they are perceived as less professional than the soldiers who work as deminers for the state, but Aki Ra describes his own trajectory of child soldier to deminer to assure potential donors that his deminers will be just as professional as any soldier.

The demining career path frequently entails a history of violence. And such histories display the horrors of shifting loyalties and the ways in which those ruptures destabilize relations. Aki Ra describes the murders he was ordered to carry out as traumatic for him. When he writes his story in English on his website, he suggests that the Khmer Rouge enforced a sense of communal accountability for murder and torture:

> Every week there would be a village meeting to decide who had been good and who had been bad. Those who had been bad, for whatever reason, would have their throats slit very slowly with palm fronds. Again the villagers were forced to cheer and clap as these people were murdered and they were taught to regard the bad people as the enemy.[3]

To survive each week, you would have to clap as your neighbor was murdered before you. The categories of good and bad in this case are depersonalized and disconnected from ideology. Friendships were destabilized not because of a lack of love, but rather because of the need to survive. Although Aki Ra is unreserved in sharing the details of his life, this was a rarity in my fieldwork. Normally, stories of survival during the Khmer Rouge and the other wars in which deminers were active leave many violent things unsaid. What was almost never discussed among deminers was the violence they committed as soldiers. There was typically a shared sense of the potential for them and their colleagues to be at war and an atmosphere of suspicion on the minefield. Histories of conscriptions suggested conflicts that were not ideological but rather about duties and hierarchy and the ways in which a person's orders would turn them against their former friends. In practice, this meant that the deminers worked with both former enemies and potential future enemies.

As with many places recovering from civil war, Cambodian minefields were full of villagers who were "intimate enemies" (Theidon 2013). That is, they are former combatants who must live "postwar" as neighbors. Relationships normally understood to be stabilized by trust, love, and shared values had been ripped asunder. All war has stories of betrayal, trauma, and violence, but in civil war, when a country's people fight against themselves in different factions, violent betrayals often take on more intimate tones. In Cambodia, I heard stories about siblings who had fought against each other, stories of marriages destroyed by mutual traumas and shared lies, and stories of survival that had demanded betrayal. These are the villagers who would be painted as innocent victims but who are no more or less deserving of a decontaminated landscape than any other human being.

Let's return to Hov, the former soldier I met in 2010 in the K5 belt. When he described his combat experience, he explained, "You had no choice, you had to fight. There was killing everywhere. Khmer killing Khmer."

"Khmer killing Khmer" was a phrase I heard repeatedly in the field.[4] It often came up when people were visiting memorials or explaining the presence of landmines. It was often the way in which members of the military narrated the history of the Khmer Rouge.

Hov was careful to elide details about his alliance in his own narratives, a silence I found in the field. His silence, as I have mentioned, likely indicated that he had fought for the Khmer Rouge. Or, perhaps, in the aftermath of civil war, his silence expressed how ideological orientation did not matter—that what side he fought for made no difference in his community since, on the ground, most villagers had been forcibly conscripted regardless of their beliefs. Such a framing of these forced conscriptions, where ideology becomes unimportant, can have troubling implications.

Indeed, many of the villagers in Hov's area were virulently anti-Vietnamese, a sentiment that has caused many problems for the Vietnamese minority population in Cambodia, but also a sentiment that has become associated with the opposition party (some would say even encouraged by opposition party leaders). Taking a racist anti-Vietnamese stance often led to a reframing of historical narratives where the ousting of the Khmer Rouge was seen as a "Vietnamese occupation," which in some tellings diminished the horror and violence of the Khmer Rouge. In the K5 belt, it is also well known that the Vietnamese-led army installed most of the landmines, leaving people with dangerous problems for which they often blame the Vietnamese (Oesterheld 2014; Hinton 1998).

In these areas, then, the understanding of the Khmer Rouge soldiers is influenced by histories of conscriptions by multiple armies, racist anti-Vietnamese sentiments, and the understanding that those who survived were "forced to fight." This is not to say that there is no stigma to being a veteran of war—civilians associated former combatants with bad karma, especially if they had been injured during the war (French 1994). Recovering from the disruptions of violence, knowing that neighbors had once been enemies, affects the overall neighborliness of a village. Even if underlying feelings of mistrust were not expressed in direct testimonies people made about each other, there were often generally understood public secrets (Taussig 1999). Eve Monique Zucker (2013) explains how one man, whom she calls Ta Kam, was known as the one all the other villagers held accountable for the deaths of their family members during the Khmer Rouge. As the village leader, the man had reported villagers to the Khmer Rouge soldiers, who then executed the people he had described as betraying the Khmer Rouge (20–21). Although the villagers insisted that no one had any problem with the former village leader, he was generally ostracized. The man himself insisted he lacked agency for the actions he committed during the Khmer Rouge (which he never directly admitted to committing). Zucker writes that "like the villagers, [Ta Kam] presents himself as a passive entity, subject to the whims and will of more powerful outsiders and circumstances" (87).

In stories like these, where humans become represented less as individual agents and more as "passive entit[ies]," anthropologists often point to Buddhism and its tenet of no-self as a system of logic to account for why violence can be understood as collective rather than individually accountable (Zucker 2013). Zucker instead looks to history and social life to suggest political reasons that people in Southeast Asia express uncertainties about violence. Zucker has pointed out that her interlocutors in a postconflict village, where fighting had continued through the 1990s, professed their kinship with former enemies. These villagers did not easily impart details about their own violent actions in the past (4). In a way, the ambiguity about actors and actions of violence can be seen as a just-in-case strategy designed to conceal how one felt about the violence of the past, whose side one had been on, and what one might be responsible for. My interlocutors, like the villagers in Zucker's ethnography, had been forcibly conscripted to opposite sides. Most of the regimes in Cambodia's twentieth-century history conscripted young people, and the Vietnam-backed regime in the eighties even forced civilians to fight guerilla warfare. Could the Cambodians I know be hiding details about violence just in case? Just in case they had fought against each other?

·   ·   ·   ·   ·

Aware of these suspicious relations on postwar ecologies and curious how nonhumans figured in these relations, I traveled to meet the other well-known companion animal that was the mine clearance industry's favored biological technology before the rats: mine clearance dogs. There I saw shadow stories exposed more clearly.

To do this, I had to get permission from the higher-ranking officers of CMAC. The only detection-dog teams available were those working with CMAC, the demining operation run by the government's military in Cambodia. They operated out of Battambang, a neighboring city of Siem Reap that attracted far fewer tourists. Battambang was a thriving community of local artists who were attracted by the famous Cambodian-French circus Phare Ponleu. The houses were old colonial mansions repainted by Khmer businesses in multicolored pastels and jewel tones. These colors overlooked a muddy Sangkar River that ran through one of the most densely mine-contaminated provinces in Cambodia.

Brahm, a deminer I had met through a long chain of previous contacts in the city, said he would drive me to the CMAC office to meet these high-ranking officers before I went to the minefield, which was about an hour and a half away from the city by car. He picked me up near one of the moneylenders at the Battambang market. He wore a polo shirt with embroidered guns where the emblem usually goes. When I was with Brahm, the mood was laced with uncertainties on both sides. We remained suspicious of each other's intentions even as Brahm seemed to be telling me directly about the corrupt practices in the demining industry.

He asked why I wanted to see the deminers, and I told him I was studying how Cambodians worked with animals. Then he said to me, "I have two jobs." He showed me his photo IDs—cheaply laminated portraits of himself in uniform. "I am both in the military and a deminer. Brahm is very important. Brahm teaches all the military at the border. Just last week I arrested two people." This confused me. I believed that Brahm was trying to impress me with this dubious claim about his "two jobs."[5] He even said, "I think your boyfriend will be jealous if you come with me to the countryside."

I did not understand how or why he thought I would find it impressive when he told me he acted as a secret police officer for the government. This was early on in my fieldwork and I was not aware of all the rumors that surrounded the mine-action industry. He said he had arrested people before they realized he was a military officer: "When I wear my demining uniform, no one knows I can arrest them. And then I do. Hun Sen himself has given me acres of land for this job—acres in Takeo and Banteay Meanchey. I will someday build hospitals there."

Here Brahm spoke of a secret within a secret. As a military officer who could arrest people, he was in disguise when he was in his demining uniform—another case where you could not trust what you saw.

Hun Sen's "gift" of land to him seemed suspect to me.[6] I had to read between the lines to understand what was going on in his narrative. When he said that Hun Sen, the prime minister, had given him acres of land, Brahm indirectly alluded to the government takeover of village lands— this was indirect because he never spoke of the people who had originally owned the land nor did he use the term *land grab*. The government has claimed that this practice has stopped, but Brahm's testimony speaks to a concrete instance of at least the rumor of the military taking over lands, reinforcing how even postwar relations shape its ecology.

What Brahm named as acres of land given to him by Hun Sen could have been land that had been grabbed by the government. As of 2012, the Cambodian state, foreign governments, and private investors had forcibly taken over 2.1 million hectares of land (more than half of the country's arable land), displacing villagers and clearing forests to make way for plantation crops (Neef and Touch 2012, 1; Beban and Work 2014, 598). The postwar ecology shows its logical extrapolation—the violence of warfare that shapes state power over land. Like the landmines, the government extended war past its end date, taking land away from villagers and giving it to high-ranking deminers. Oftentimes, these lands were grabbed because villagers who had survived the war feared the state even with no direct threat from it. Just a few weeks before, Cambodian military police officers had shot dead environmental activists at the border and had arrested a foreign-national activist. Was Brahm implicated in this attack and arrest?

Because of my shock at Brahm's admission of being both a soldier and a deminer, I was mostly silent in the car ride from Battambang to the government headquarters of CMAC. I examined his ID card more closely, a picture of him in uniform in a little plastic card sleeve. I wondered if I should have gotten in the car with him. On the other side of the sleeve was a picture of Buddha. When I pointed at it, he said, "That keeps me safe." The card had two things that kept him safe, actually: the military ID that gave him government protection and the Buddha that kept him spiritually blessed. While he only pointed at the Buddha, his talk about the state and the acres Hun Sen himself had given him transitioned easily to a conversation about spirits.

He pointed at a triangle with a handwritten filigree of a Buddhist blessing in felt marker on the car ceiling as a form of protection. Yantra spells are sacred scripts and shapes, magical charms drawn by *krau khmer* (animist shamans) or Buddhist monks that grant powers, protection, or charm to the bespelled (Cummings 2012, 48). Retrospectively, the yantra above our heads was an ominous foreshadowing of the parallel state and spiritual surveillance I encountered during my fieldwork and of what I began to suspect about Brahm. I learned how to mistrust from my encounter with him.

We pulled into DU2, the state organization's mine-action headquarters outside Battambang. We walked by a cement table outside where various deminers sat. A handsome older man with a shock of white hair smiled and nodded at Brahm. A woman smiled at me as she left on her motorbike. Brahm led me into the building through heavy wooden doors. Downstairs inside the blue tiled building, we met with the deputy director, and upstairs we met with the chief director. Brahm hastily introduced me to the officers, telling them that I would see the minefield and that I was a researcher. He explained to them that I knew Khmer and that I came here the other day (or that I would come again—the words he used made the timing ambiguous). Then we walked back down the stairs. The whole meeting lasted less than three minutes.

Back in the car, he explained to me that I shouldn't contact anyone at the headquarters, and that if I wanted to see the deminers, I should just contact him directly now that I had already met the officers.

"Don't tell the head guys we will go Monday, Tuesday, or Wednesday—because I am busy and you are busy and so we don't know for sure. Just say you will come for the visit another day. Maybe next week. It is hard to say—you are busy and I am busy."

After calling many times and several days later, I was finally allowed to visit an active minefield being cleared by mine-detection dogs, but not through Brahm. Brahm returned my call only to invite me to the country-side with him for the weekend, an offer I refused because I found his intentions suspect. Instead I met up with a friend of a friend who knew a "big boss" and somehow managed to get me a ride to a mountain range near the closest minefield. He told me I would meet Brahm there—I called ahead but Brahm's phone had a message saying it was disconnected.

The two drivers were quiet in the front seat once I spoke a few Khmer words. We drove through tapioca crops to Mount Banon, where the morning team worked. As the road got bumpier, the man in the passenger seat got directions from someone to go all the way to a bridge. From the rearview mirror swung a little frog made of red and golden thread, meant to bless the car. On its side mirrors was more red string, whipping and hitting the windows.

We left very early, which gave me time to watch the changing countryside from semi-urban areas to farm crops, rice fields, and tapioca fields. The tapioca had tall white stalks, and the ground beneath, when burned, was blackened and littered with white specks. We drove on a dirt road with massive holes that unbalanced the car. Not many people farmed or walked around in this part of the countryside. I saw a young woman dressed in pink with a straw hat who held the hand of a little boy in the middle of a recently burned tapioca crop. An old lady squatted by her house while another old woman crouched on a raised platform and rocked a child back and forth on a green hammock. Finally, around 9:20 a.m. we turned onto a narrow raised pile of dirt, where the driver stopped and let pass a giant farm wagon with a completely flat bed and a single seat in front. We came upon a clearing in the tapioca crops and a rectangle of water where Brahm, in uniform, smiled and waved. He wore a pink and black checkered scarf, a *krama*, traditional for Cambodians, who use them as a towel to wipe off sweat in the heat.

Brahm had me sit before a table and a white board with a polygon map of the minefield. He began an introduction to the minefield, with his name and position, and described his team of ten staff—a diagram showed his photo at the top of a pyramid with eight deminers and two brush cutters below him. Two of the deminers were dog handlers with portraits of themselves with their dogs: one was a woman, Nuon, and her dog, Costa, and the other a man, Sina, who had a dog also named Sina.

Brahm said they had found five mines in the area so far and that they would stay for a week or so more and then move onto other minefields. He recited details about the village: the population was about 500 people with 135 families; its name was Svay Sau (White Mango).

Then he allowed me onto the minefield, but first he had me put on the Personal Protective Equipment, which everyone, even in Khmer, called the PPE. I had to wear a thick vest and a hard hat with a plastic visor that came down in front of my face. The vest and helmet were standardized according to International Mine Action Standards set by the Geneva International Centre for Humanitarian Demining. The vest is officially called "body armor" and made of Kevlar to protect a person's most vulnerable parts from antipersonnel landmines. They are not guaranteed to protect people from all scenarios of explosions (Smith et al. 2017). As we walked over upturned dirt and fallen grass stalks, I attempted to make a videorecording.

On the field, two bright red umbrellas with woven bamboo mats and water bottles were planted at angles toward the sun. Brahm let me follow him into the minefield but did not stay directly in front of me. I could not see peripherally because of the helmet. At one point I stumbled under the weight of the vest. I followed Sina and Sina at first and then Nuon and Costa. Every time we came near one of the dogs, Brahm murmured their names low with encouragement. The dogs sniffed at the ground but often sat and looked up at their handlers with eyes open and a long, panting tongue.

They were sleek black speckled dogs with pointy ears. Their coats shone. The handlers explained to me that they were very tired and needed a break because they had been working since 6:00 a.m.

As we approached the dogs, Brahm cautioned that I should approach the handler first and then the dog because the dog might think I was a strange person. The dog would not trust me until the handler greeted me. For a little while I kept my distance, but the dogs seemed unfazed by me. They just sat under the umbrellas, panting, giving none of the signals dogs do when they are nervous, defensive, or about to attack—all signals I had expected after reading a description of the mine-detection dogs' behavior when meeting a stranger (Kirk 2014). The promotional materials on mine-detection dogs emphasize that they are a loyal Shepherd breed that is as defensive as it is intelligent, a disposition that correlates to the militarized practices in mine action. Militarism has certainly influenced training, and its practice as well as its aesthetic has influenced canine breed selection. For example, an Australian trainer has successfully trained Labrador retrievers as mine-detection dogs, but they have failed to get traction perhaps because floppy-eared Labs have a different aesthetic than Shepherds with their wolf-like looks, pointy ears and long nose. Another breed, the Swedish Drever, which resembles a Bassett Hound with its patches and floppy ears, has also been tested but has not been as popular (MacDonald et al. 2003). The animal studies scholar Robert Kirk (2014) has described a transition from Labrador use to Shepherd as a transition in the understanding of the "intersubjectivity" between handler and dog. A Shepherd's wolf aesthetic stresses the "wild" and aggressive nature of Shepherd. Later, I will see how this contrasts with the ways in which the rats are promoted as friendly cartoon heroes and as affectionate pets. Perhaps this contrasting marketing was an overcompensation for the stigma associated with rats, but it also had the side effect of pushing back against war materialities. The military-waste-detecting dogs, though, were almost as untrustworthy as the ground they smelled, as I discovered when I met them.

I followed Brahm's instructions, saying hello to Nuon, the handler, first. She smiled at me.

"Do you like Costa?" I asked in Khmer.

She and Brahm both corrected me: "I *love* Costa."

I looked at Costa and said that Costa also loved Nuon. They both smiled and agreed.

I spent no more than fifteen minutes on the minefield before I was ushered away by Brahm. The deminers followed us. He kept saying that soon they would break, at 11:30 a.m., for lunch. But it was not near 11:30 a.m. After fifteen minutes he said finally in English, "Okay, now we can go."

I told Brahm that I would like to visit on Wednesday as well. Brahm said it was too close to New Year—there wasn't enough time for a visit.

Brahm presented me with a sign-in and sign-out sheet. "I forgot," he said.

"Okay," I said, "so I arrived at 9:20 a.m. and I am leaving at 10:50." But Brahm stopped me from writing and pointed at the sign-out time and said, "Put 12 o'clock."

"What time is it?" I asked. And he just repeated, "Put 12 here."

I wrote "12 noon" even though it was only 10:50 a.m.

I never saw those people again. Instead, I focused on attending to the rats, who were more interesting to me, anyway, with their blind eyes and shivering noses. I later told Hien about my experience with the dogs. Hien was the rat handler and platoon supervisor, and he laughed at the part of the story about Brahm and my visit to the tapioca minefield.

"Yes, because," he said, with a glint in his eye, putting a finger to a smiling mouth, "the dogs, they don't work."

"They don't work?"

We were watching some mine-detection dogs because there was an exposition that day for the state demining organization at the training quarters. Representatives of mine-detection organizations from all over the world as well as funders of these organizations had arrived to see presentations and to visit nearby minefields. Hien's smile widened as he saw my shocked reaction. "How do you mean they don't work?" I asked.

"I never understood until I worked with the dogs. Because, you remember, I was a dog handler before this job."

"Yes," I said. In front of us, a Malinois with a long neck was following the CMAC and Norwegian People's Aid handler in the tall grass nearby.

"You see, even now, that dog is not approaching his handler. She is blowing her whistle and he is supposed to sit right beside her, but he does not."

The dog looked up at his handler but, it was true, he did not come when he was called. The dog wore a special vest with the names of all the organizations that sponsored his training. The handler and the dog, I assumed, were rehearsing for the exposition that would happen later in the afternoon.

"When I worked as a deminer with a metal detector, I was in a minefield with my friend who worked with the dogs," Hien continued. "We had a dog there. We were far up north in the jungle. It was very dangerous. But we had no equipment. We got there ahead of the equipment. 'It's okay,' I said to my friend. 'We have the dog! Just use the dog.'"

"'No, no, no,' said my friend. 'We must wait until we have the metal detector, too.'"

"'But why?'"

Hien paused and we continued to watch the dog, who was now tired in the brilliant white light of the 10:00 a.m. sun. "He did not tell me why," he said, mimicking the silence of his friend with his arms crossed and a frown. "So, we had to wait until the equipment arrived. Then once I became a dog handler, I understood."

Hien didn't elaborate.

His silence made me realize that there was a shadow beneath the story. The dogs didn't work. Corruption had meant that the funds directed toward their training, with necessary refresher courses, would be used as bribes and given to people within the military. Later, other people corroborated some of these rumors. People told me about military police disguised as deminers, similar to Brahm's story. I was even told that villagers probably wouldn't assume that a deminer was in their village to clear landmines but rather, to conduct government business with a hint of state violence. They confirmed Hien's suggestion that the dogs "probably didn't work." The rumors made me think about Brahm's two jobs. I wondered about this as a mine-action shadow story: a police officer dressed as a humanitarian deminer, benefitting from government land grabbing and arresting activists. Had Brahm lied to me when he said the dogs needed to rest because they had been working for four hours as they sat in the shade panting? Or were the dogs just for show and was the money for their training being redirected to more nefarious border-patrolling violence? Did Brahm have two jobs? Or maybe it was Hien, angry at the government, who was lying.

I couldn't be sure. This uncertainty about relations paralleled the uncertainty about explosive landscapes. The untrustworthiness of humans and even dogs, I was learning, was part of the postwar ecology.

·   ·   ·   ·   ·

On a demining platoon, where most everyone has had combat experience, these ruptured relations are amplified by the bombs and the military ethos. The labor itself is choreographed around individuals who must protect themselves from a potential explosion. Typical mine clearance is characterized by silence. One by one, deminers work in a long, uniform row. They wear bulletproof aprons that extend past their knees. They wear heavy helmets with a long visor. They stoop over the land they clear, intently observing the extended metal detector before them and walking slowly. When they tell their stories of violence and conflict, they tell them in shadow stories, which means that you cannot trust what you hear.

The deminers on the landmine-detection rat platoon had a range of ages, from twenty-four to fifty-four. As soon as they began their training, there were visible divides.

The rat handlers, all experienced deminers, had to be hired after the arrival of the rats. There were a few managers and the Tanzanian consultants to lead the rat-handling training. The NGO planned on hiring two of the villagers who had helped clear the grounds and cut the grass of the training fields. The rest would be interviewed in the month after the rats' arrival. Half the deminers would be employed by the NGO and half by the government agency CMAC. The bureaucratic tension was palpable in the building already. Hien had been hired as a supervisor, and he recommended his older brother, Chann, to be hired to supervise the field training. Chann constantly criticized the government but would tell me he was worried about his brother because he was too outspoken in his criticisms of the government. All the Cambodians I knew had moved to the city of Siem Reap or to the building with the rats while their families lived in villages, some of them quite far away. Much of our daily small talk focused on how their wives were and how their children were. They would show me photos of their young children and wives on their phones. This was standard for deminers, who, like soldiers on patrol, had to live far away from their families for months at a time.

Before the rest of the team was hired, Liz, the British manager, and Chann discussed how to arrange the sleeping quarters in the two rooms adjacent to the rat kennels. The deminers were expected to sleep in the buildings for the year as they trained to be rat handlers. After the rat technique was certified by the national certification program, they would be deployed in platoons throughout the countryside and sleep in guesthouses or sometimes in camps in the rural minefields.

"They'll probably all need access to a guesthouse with air conditioning when they get deployed," Brad told me. "They'll have to travel with the rats and work with them earlier than other deminers because the rats are nocturnal, they don't like working in the day."

Already, the rats were changing the normal methods of landmine detection. But Liz was first concerned with getting certification and managing the new employees. "We have two rooms and so we can separate the men and others. One men's bathroom and one women's bathroom."

Chann said, "Okay, yes. It depends on how they want to be separated."

When I heard that, I didn't yet realize it, but Chann was referring to potential tensions among the employees within the group. As I've mentioned, deminers have backgrounds of being enemy combatants, usually soldiers drafted to one side or the other, but they also have tensions when it comes to the competition between minefield-clearing organizations for minefields to clear. These tensions became apparent when the workers decided to sleep separately based not on gender, but based on the organizations that employed them. The two women and the nonbinary person were government workers who decided to set up their cots in the same room as their fellow employees, while the NGO employees were set up in the other room. At meetings, this institutional divide made its way into the first few weekly debriefings the deminers had with Liz.

"How much money do the NGO workers make?"

"They make the same as you, mate," Liz said. She shook her head slightly at the translated words since it was a question that had been repeated several times in several different ways. The problem was that the NGO workers' weekly salaries included money for lunch, whereas the government workers received this allowance as pooled cash. This led to grumbling among the government employees. Their pay had also been delayed for bureaucratic reasons. They did not quite trust her answers.

The division between the teams was apparent in their uniforms. NGO workers, even the supervisors, wore soft cotton T-shirts with the NGO's logo on them, a stylized rat in silhouette with three circles representing landmines. In contrast, the men and women of CMAC in the rest of the building wore standard military uniforms with gold tassels on their shoulders. One deminer joked with me about a military-inspired cap I wore to block the sun, saying that I was a high-ranking general. The NGO solved this problem by giving me a hat with the APOPO logo. Non-Cambodian NGO management frequently expressed annoyance that their Cambodian workers, even the NGO staff, noticed and remarked on the difference in the quality of buttons on the uniforms of higher-ranking officers. APOPO's uniforms—tan cargo pants and cargo shirts with simple baseball caps—contrasted with those of the CMAC deminers,' with their military hats with crests, fancy buttons, and golden embroidery for officers.

The workers were not only divided along institutional lines; they were also divided along lines that were rarely spoken about: political and historical. These were shadow stories. Because of their fears of surveillance from suspected enemies, I have disguised the ways I represent the stories under their approval and advice.

In the mornings, I rode my bike through the darkness to the Siem Reap office. I met the APOPO van before the sun rose to accompany the Tanzanian consultants on their way to the training field, a forty-five-minute car ride over unpaved roads and through villages wealthy from the economic opportunities that tourism had brought to Siem Reap. The saturated colors of the sun painted palm tree silhouettes over dirt and an endless horizon of rice fields.

The car we used to drive to the minefield was a safe place, according to Chann. Chann lived in the city and commuted to the village where the training grounds were. Even though he was employed by APOPO, he borrowed the government car, a junky jeep with a CMAC logo, covered in red mud from the unpaved roads. This state-sponsored logo offered some security. On village roads we wouldn't get stopped by police for illegal tolls. But this was not the kind of security Chann was talking about.

"You can only say these things in the car," he would say when he told me stories about the current government's injustice. Chann had many complaints about the government and how he even suspected it of violence, but he told me so only in the car. In the minefield, he told me, he kept silent because he worried the government employees would act as informants.[7] Plus, he thought he was safe to say what he wished in the car. Maybe the car afforded all the people in it an aura of trustworthiness.

Chann was a supervisor at the demining unit, and he managed his younger brother, Hien, a tall, skinny man who was a lower-ranked NGO supervisor on the minefield. They had both grown up in a refugee camp in

Thailand. In this regard, they were unique among the team, as most of the other deminers their age had been soldiers during the Khmer Rouge era and the 1990s. Chann looked like a bulkier version of his younger brother, both had a rectangular jaw and spectacles. Chann used to be a champion boxer and he carried himself a bit more heavily, shoulders always slightly slumped forward. After I got to know him better, I imagined that his worries about spies and his brother, and not only his bulk, burdened him.

Every day at 4:30 a.m. Chann drove the Tanzanian consultants and me to the village area where the mine-detection rats were being trained to detect TNT. The new team of deminers were also being trained to handle the rats by the Tanzanian consultants. Hien, Chann's brother, did not need to commute because he lived with the deminers and the rats in the CMAC compound near the village. The deminers were divided into two teams, five government employees and five NGO employees, but this did not include the CMAC supervisors who visited the compound and the CMAC security guards who lived on the compound. Hien was surrounded by government employees, which was why Chann worried about him. He thought it was dangerous when his younger brother criticized the government around government employees.

"They might be spies," he explained in the car.

"[My brother] says things like he doesn't think they [the government people] are good. He says things like if they [the government people] controlled the money, the deminers would never get paid."

"But," I asked, "is it true? Hien said that the deminers would never get paid if CMAC [the government organization] was in charge of the budgets."

"Yes, yes," Chann said, "it is true." Then he laughed. "But you can only say such things in the car! You can't say such things in the field where people are watching."

Then he told me stories of how the government "disappears" people. He elaborated on how some of his neighbors had said bad things about the government and the next week they had been found dead. Someone, and here he implied the government, had poisoned the rain barrel where the family collected cooking and cleaning water. The whole family had been killed. Chann seemed to think it was the only logical explanation for his neighbors' murder. I asked Chann how he could be sure it was the government.

Chann insisted to me that his neighbors had criticized the government's ruling party, the Cambodian People's Party (CPP). The week after that, they had been poisoned. He also expressed this as a present reality of how governance worked in Cambodia. For him, distrust by Cambodia's people was justified. A village chief had tried to cheat his in-laws out of some land, for example. It was a common practice, he said, for official authorities to pretend that their land claims were larger than they were and to evict poorer villagers, part of an epidemic of land grabbing in Southeast Asia.

Nearly every day, Chann repeated to me that we could discuss "such things" only in the car. The car was a haven where he could vent to us about the dangers of the CPP government, the spies, and the corruption he saw every day. While the Tanzanian consultants just listened and nodded off in the back seats during those early mornings, I sat up front, forcing myself to remain alert. In the car, that junky rumbling jeep, Chann felt free to express himself in ways he didn't in the minefield. There everyone, except for Chann's brother, Hien, usually remained silent, focused, I thought, on the tedious and dangerous task of finding landmines.

One day all of this changed. Chann was quiet when we drove to the village. As we neared the training ground, he said, "Bunthy told me yesterday that CMAC would like to take the car every Friday for checkups."

CMAC was the state organization—these were the employees whom Chann feared as spies. However, to me Chann's silence didn't seem warranted by CMAC's weekly car-checkup plan—the car had a lot of problems. Once, it overheated and we needed to wait in a gas station for hours until it cooled. It sometimes didn't start.

"Oh, okay," I said, still curious about his silence.

Chann looked straight ahead as he said, "I don't know why they need to check the car. We have that in our NGO budget so they don't need to check the car. I think that they are not checking the car."

Chann didn't elaborate, but he meant that we could not speak freely in the car anymore. It was considered bugged. This was another shadow story—the potential of surveillance, and through that surveillance, the potential for violence. Chann speculated that "they are not checking the car." Behind this example of state surveillance was a threat of violence—a potential to be "disappeared" or poisoned like Chann's neighbors.

But I was more delayed in understanding that Chann was warning me. Right after Chann explained that he "didn't know why they were checking the car," I asked, "You mean they've been recording us the whole time?"

I heard the clumsy worry in my own voice. Chann did not answer. The Tanzanian consultants were napping in the back. We parked at the minefield headquarters with a soothing sound of tires on sand. The work day began and we never spoke of the bug again. From then on, my rides to the minefield were as silent as the mine-detection work. The car became a lingering shadow to my mornings which, once the humans and rats became friendlier with each other, was a daily reminder that the rats did not resolve all the problems of fear.

.   .   .   .   .

One morning, Chann and I had breakfast together before going to the demining unit. This was a break from our routine. Usually, we reached the dirt-road part of the highway in the Roluos village center and avoided stopping for bread because it always took so long.

We sat at a café, and I dipped bread into my sweet coffee while Chann ate his noodles.

"I am very tired. I could not sleep last night," he said.

"Oh, no, I'm sorry. Why? Are you stressed?"

"Yes, very stressed. But I think the rats are doing very well," Chann said, "but we will stop the training of two or three rats because they are not ready. We don't want the CMAC consultant to make bad reports."

A consultant from CMAC had arrived the week before to check up on the training of the rats in the minefield. The platoon did not operate as normal under the watchful eyes of the CMAC consultant. This was Chann's explanation for why we were having breakfast and would stop the training that morning. Some people, including CMAC employees, had told me that they were worried the CMAC consultant was a spy trying to steal landmine-detection rat techniques or sabotage APOPO's national certification.

"We have to be very careful because CMAC is watching us. They have spies everywhere."

"Yes," I said, because this was something Chann has told me before. I responded with what I thought was a joke, "But I'm sure you have your spies, too."

Chann laughed a lot at this.

"You know," he said with a wide grin, "when you said I have a spy, you are right."

"Oh?"

"Yes!" Chann rubbed with hands together. "And I know that all the deminers drink every night. Even though we told them not to. I know because my spy tells me these things. " Chann paused and leaned toward me. In one breath he said the name of the spy and "you cannot tell anyone that he is my spy! I trust him."

Espionage shadows trust—trust is necessary only when some are untrustworthy. Chann told me then that he thought he needed a spy because he distrusted the ruling party, the Cambodian People's Party (CPP). Although Chann was afraid the CPP was spying on the landmine detection with rats, he was quite open to people about his support for the opposition party and made it clear that he did not support the CPP.

The CPP has been in power since 1992. There is much tension between those who support the ruling party and those who support the opposition party (Nam 2011). Some people who supported the opposition party suggested to me that the party is a holdover from the Khmer Rouge. Or, in another critique that seems to contradict this but also comes up in the same conversations, they suggest the ruling party is aligned with the Vietnamese state. Thus, the platoon's two teams became stand-ins for old complaints, like Chann and Hien who hated the institution of CMAC because they saw it as a continuation of the Khmer Rouge leadership, since the CPP included former Khmer Rouge leaders.

So the CMAC consultant spied for the government, and Chann had someone spying on CMAC. These suspicions of espionage evoke government rule during Cambodia's civil wars, and the former military enmities overlapped with the institutional divides within the platoon. Historians have outlined this constant suspicion of spies as "the panoptic principle of rule" (Bultmann 2015, 99) and a "hypervigilance" (LeVine 2010: 10) that led to a prescribed set of behaviors "in which everyone acts *as if* being constantly watched" (Bultmann 2015, 99; original emphasis). Moreover, surveillance was incentivized with labor relief, extra food rations, or safety from torture. This is why the Khmer Rouge soldiers reminded Cambodians that their organization, the Angkar, had "the eyes of a pineapple," which referred to the multiple segments on a pineapple's shell and symbolized how many spies were enforcing its strict labor and communist rules (Bultmann 2012). These habits surrounding surveillance continued in behaviors and anxieties during my fieldwork.

Although the surveillance Chann encountered and led were likely cases of corporate espionage, the stakes were higher than corporate espionage because stories implied potential violence. Some deminers feared that the government had poisoned their neighbors or, as people on other platoons told me, government deminers doubled as undercover military police officers who arrested environmental activists.

At one meeting we gathered in a circle. Moch, a CMAC deminer, had been promoted in her tasks to be someone who double-recorded the rat's behaviors in the pits; normally, the deminers recorded the behaviors afterward but Chann decided that Moch and her demining partner should record in real time. Chann had not said that people misrecorded the scores, but I think that he trusted Moch to record things correctly.

One of the other deminers took issue with this. In fact, he was the person Chann had told me was his spy. He pointed at Moch and said, "I'm not sure why we have to have people observe us like this."

.   .   .   .   .

Moch would often invite me to sit with her and share coffee. We were lying down on Sovannah's cot together, looking at the internet on her cell phone. She showed me photographs of her brother on Facebook. A skinny man stood next to his wife in a bridal photo, their clothes designed to look like those of royalty, with gilt brocade and sashes of traditional dancers.

"How many siblings do you have, sister?" she asked.

"Four," I told her. "How many do you have?"

She showed me her brother's Facebook on her phone and then she said, "This is one and I also have an older sister. I am the youngest."

"What is it like to be the youngest?"

Moch shrugged again and said, "I was also told stories about my sister. That I am lucky for being the youngest." Moch was in her thirties, so she had few memories of the wars that had occurred in her youth. Her father had been a soldier, and in the nineties there was still fighting at the border of Thailand. Moch's stories were often retold stories from her parents— from her father, who protected the border at Preah Vihear, or from her mother, who talked about her siblings.

"My brother had been kidnapped, my mother told me. And my sister was thirteen. They had him trapped in a house. Then, my mother sent my sister to rescue him."

She looked away from me, briefly clicking on her brother's photos to find a photo of her sister. We lay on our stomachs in silence as we continued clicking through her siblings' photos. A few more flashes of a woman and man smiling at the camera lens.

"The soldiers let my sister and brother come back," she said, finally.

"Why did they do that?"

Moch showed me a photo of her sister's wedding. Like a lot of Cambodian wedding photography, it showed a woman painted in pale face makeup with elaborate false eyelashes and a crown of ornate hair woven into costume jewelry.

"I don't know . . ." She trailed off and then added, "My mother never told me. Not my sister, either. My sister never talks about it. They just say that I am lucky to be the youngest."

The violence of the past was left to our imagination, and what her sister went through was a shadow story. Her mother and sister were silent about this violence. The soldiers mentioned in the story were never called Khmer Rouge soldiers. Like Chann, Moch didn't clarify which soldiers. The soldiers were just a figure that could belong to any side. This kind of mutability was also important to shadow stories. Anyone could be a violent person.

·   ·   ·   ·   ·

Most of the time, the minefield was mostly silent, interrupted only with the distant sound of Buddhist temple bells and the monks' chants that greeted the dawn. In the neighborhood though, I could not sleep during the first few nights in my wooden house. I lived in the darkened top part of a structure built on stilts. In the villages, the stilts usually frame an out-door room where families sleep under mosquito nets to enjoy the evening coolness in good weather. But instead of empty space, the bottom of my place had been renovated into a concrete-tile house where the Khmer-Japanese owners lived. In my wooden upper home, the ceilings were high and the open windows and back and front doors allowed for a cross breeze even on the hottest days. The cross breeze and the wooden beams meant that I lived inside-outside—bats flew into the house, rhinoceros beetles got swatted in from the porch by the downstairs cat, and a family of mice leapt from beam to beam to my four-post bed, sometimes waking me with their squeaks and the sound of their scrabbling paws. Through the cracks in my wooden walls came a symphony of insect and other night crawler sounds. It became easy to imagine monsters. One creature made belching and sucking sounds in the mud in the surrounding jungle area, like a huge beast with a wet mouth. Two insects signaled to each other in high-pitched violin-strum noises. Surrounding all that were the chatter of cicadas and the howling trumpets of the yellow dogs that roamed the streets and homes. The demining rats were part of this milieu of sound.

The contrasting silence during the day was an anticipatory quiet. It did not feel calm, despite the tones of bells and chants as the sun rose red-gold. The deminers were working hard to get their national certification, and they focused on learning their new task. One member of the pair of humans would gather a rat from the transport cage under a grove of trees and, her footsteps crunching gravel in the cleared path, walk the rat to the mine pit the handler's partner had lined with measuring tape. Then the handler would take off the metal grill on top of the plastic transport cage, the sound of metal clanging loudly in the quiet. When putting on the har-ness, the handler would murmur to calm the rat as she held him, stroking him softly while clasping the plastic buckles. The sound one could hear was mostly the swishing of the rope as the pair of humans pulled it to guide the rat in a straight line across the pit. Sometimes they would laugh at the rat; sometimes they would murmur about how slow he was being; sometimes they would joke with each other. Mostly, though, in the early morning, we all watched silently as the rats explored the earth beneath their paws.

The rats were almost always silent, at least for human hearing, save at night. At night they would squeak to each other, "talking and talking and talking," the deminers called it. "So much gossip, just like Cambodians," someone told me once. Most of the sounds the rats made were ultrasonic. We humans could not hear their chitters and conversations about the world around them or even know if they had them. But we could imagine their sounds from the ways they moved in the world. No rat was ever distressed enough to squeak loud enough for humans to hear, except for one rat, Merry, who squeaked loudly when he had dreams. The humans used this to relate to him, too.

"Sometimes," Chann said, "I remember bad things like Merry."

The rat sounds that were within human-hearing range allowed humans to relate to them. Interpreting their squeaks gave people a way to acknowledge that the rats were fragile and worthy of trust, perhaps even love. But more than that, the sounds the rats made in the human-audible range were ways that the humans could acknowledge person-like behaviors such as "gossip" or "fear." The sounds of rats indicated to the humans that communication could traverse species lines. This became important as it paralleled ways of traversing shadow stories and political-institutional tensions in the platoon.

3 Even the Clouds Lie

We often finished the work week with a party. The supervisors would enlist one of their wives to cook, and we would bring beers, and all the deminers and the supervisors would have a meal together.

At one of these parties, I looked up at the clouds. "I think it'll rain," I said.

"No, it will not." Chann laughed. "If it rains, I bet you. Five beers." And he splayed out his fingers to stress the five.

"Okay, but the breeze is coming. And there are clouds in the sky. How do you know it will not rain?"

"Because," he said, "in Cambodia, the clouds lie. Everything is a lie."

Most people were ignoring our conversation and continuing with their appetizers of duck meat and taro chips, but people laughed at that. Lies feature in a lot of jokes—puns, lies, and double meanings are all part of talking to coworkers. Multiple people had told me that they suspected their coworkers to be spies. Everyone may be a liar.

It did not rain and I still owe Chann his beers. This makes me a liar, too, I suppose.

I do not want to consider the double meanings and unspoken messages of people's words as mere lies, as Chann suggested, but rather as portrayals of both fear and love on the minefield. In the chapter that follows, I explore these affective contexts, attending especially to their correlated technology-actors: bombs and rats. It is possible to connect the lying clouds to the ways in which a minefield also rendered relations untrust-worthy through its uncertain materialities. The previous chapter's stories depict how ambiguous expressions of agency align with military tensions in a postwar ecology, and this chapter considers what it means to be acting within a postwar ecology of uncertainties and untrustworthiness all around you. This means attending to the nonhumans and the ecology as a set of relations after war's ostensible endpoint.

The rats become key to understanding the minefield as a *postwar* ecol-ogy because they themselves are produced only after war ostensibly ends, despite its intractability, and they have been produced with the specific hope of changing a militarized ecology. This chapter considers who the rats are and how they push back against the totalizations of war's shape, which configures most human relations in the postwar ecology.

You find landmines in places that inspire peacefulness. As we learned earlier, minefields are often in a place that the soldier assumed would appeal to humans—a place with shade, a body of water, scenery—these were of utmost importance to the decision process for planting bombs. This is how the invisibility of landmines rendered signs of peacefulness untrustworthy. When deminers could not trust their own senses, they entered a place where everything else also became suspect.

The term *postwar* is a misnomer because the temporalities of wars—especially in places where military waste like explosives, chemicals, and other war materialities extend the period of warfare long past its official diplomatic end (Zani 2019). I want to add *post-* to *war* to mark the official end to better depict the war's *extension* beyond human markers of peace. I integrate the term to consider human agency when it comes to war's effects on the relations within an ecosystem. Thus I do not mean to indi-cate that war ends when humans declare it to be over, but rather I want to emphasize that war lives on beyond this point and consider how the actors in this ecology, both human and nonhuman, (re)configure it. Postwar, in this frame, is aspirational.

I employ the word *ecology* in terms of understanding *postwar ecologies* to refer to the logos of relations that extend beyond anthropocentrism—

that is, how people understand their human-nonhuman relations, which includes the agency of nonhumans to shape such logics. The scholarship I refer to for thinking through nonhuman-human relations utilizes a variety of terms including "multispecies," "interspecies," "transspecies," or even "the more-than-human" as concepts to denote a sense of thinking that includes beings and relations such as spirits, flora, and fauna (Haraway 2003; Rose et al 2012; Kohn 2013; Parreñas 2018; Khayyat 2022). All these terms, since they are legacies of Western supremacist histories of categorizations, have drawbacks, not just because of their politics but also because of how embedded in a certain categorical order they are. For instance, *species* as a term alludes to the Linnaean idea of taxonomical species; the term *more-than-human* originated in nineteenth-century Europe and North America, where prey animals were morally good and predators immoral (Benson 2013). I have chosen *ecologies* for its references to *logos* (the knowledge of) and so that I can ground its details in the specifics of my field site, but, in the end, I seek to "confound the western animacy hierarchy" (TallBear 2017) with the stories I present.

An emergent group of scholars has sought to consider ecologies in the aftermath of war to better understand warfare and military industries as integral to shaping relations within an ecosystem, often in surprising ways. The anthropologist Eleana Kim (2014, 2022) has considered the demilitarized zone (DMZ) between North and South Korea as a "militarized nature" that produces "unexpected ecological sanctuaries" for endangered red cranes to flourish, which, in turn, affect how human residents think of the landmines that litter the DMZ. These ecologies, altered postwar, influence human relations and concepts. Landmines and other forms of military waste render the "landscape as weapon," as Munira Khayyat (2022) points out, even as multispecies residents continue to resist these "technologies of death" (137). Other anthropologists have considered the ways in which decomposition from chemical warfare produces novel ideas about hope, the ways in which nonhumans come to stand for legacies of war, how war structures landscapes in places far from where it occurred, such as the suburban gardens of the United States or the war games and relationships in rural US states, and the multitude of ways in which postwar beings (humans and nonhumans) both transform and are transformed by war (Lyons 2020; Stone 2022; Crane 2023; Ruiz-Serna 2023;

Pardo Pedraza 2023; Pinto-Garcia 2022). In these ecologies, war and its aftermath configure and reconfigure relationships among humans and nonhumans.

In the context of Cambodia, war has altered ecologies in ways that are evident and in ways that are more insidious. From the colonial era, when French governments imposed extensive agricultural exploitation practices, to the genocidal regime of the Khmer Rouge from 1975 to 1979 that burned down forests and forced extractive farming practices in a time of famine, war has left its mark on the ecological relationships humans are part of. In the rural areas, people live amidst the persistence of minefields and chemicals on people's bodies from Agent Orange (Bultmann 2012; Hammond and Schecter 2012; Taksdal 2011; Kiernan 2004; Au 2011). In ways more subtle, the civil war and imperial conflicts have left the country with the infrastructure of a government that benefits from and even partners in land grabbing, which itself shapes the ecology as postwar by funding exploitative farming and other developments (Beban and Schoenberger 2019). The environmental damage reflects the damage done to human relationships. Postwar ecologies can be seen in the vulnerability of landowners, who fear signing papers because of the genocidal regimes they lived through. They end up not signing papers that legally proclaim their ownership, which makes their property ownership more vulnerable to corporate land grabs, resulting in destructive industrial projects that impact mangroves, forests, and villages (Chhim 2012).

Because human sight is incapable of assessing the presence of landmines, rendering them an invisible threat, the landmine-detection rats stepped in a with a new kind of sight. To understand the ways in which they offered a new way for humans to sense hidden dangers, it is also necessary to understand how untrustworthiness was explicitly linked—through jokes and comments like Chann's about lying clouds—to interpersonal dynamics and corruption of the political system. *Do not trust what you see,* but landmine-detection rats allowed the deminers to see things in new ways, and not only through the ways in which they detected unseen explosives. Before they arrived, however, the materialities of landmine detection reinforced militarism and untrustworthiness.

Deminers' feet touch uncertain ground. You see them beyond the bright red signs featuring skull and crossbones that warn walkers to stay away. In the demining choreography before the advent of landmine detection with rats, they worked past these skull and crossbones markers while wearing helmets and masks and aprons of bulletproof materials. Their walks were slow and tedious. Usually, they used a metal detector to sweep the ground in front of them, the sun reflecting off their long plastic visors.

"It is hard to get used to the PPE," Hien, a demining supervisor with the landmine-detection rat platoon, told me when I interviewed him during a minefield visit. "At first, when I wore it, my breath made the visor foggy and as soon as I put on all the PPE, I began sweating. The heat made me a little dizzy. It was hard for me to just go for a walk. Then, with the metal detector. It is even more difficult to sweep a heavy tool in front of you that detects metal beneath the ground."

After Hien detected a bomb, he would tell his platoon leader. He would painstakingly map each beep on a little sheet of paper so that a detonation team could come and detonate it.[1] I thought of the amputees I had met as I walked on the minefield after him and wondered how a bomb that could rip through flesh and bone could be withstood by these protections.

The PPE enhances humans in a cyborgian sense but also dulls their sensory capacity. You cannot feel the ground through the boots or gloves. Instead, the landmine detection technology, such as a metal detector, a dog, or a rat, allows you to perceive the invisible landmines. While a human's touch is limited by the PPE, postwar technologies compensate in ways that expand the body's reach in much the same way Gregory Bateson (2000) describes a blind man who uses a stick to see. To consider the stick as a way to see also extends the boundaries of what it means to have a human body, just as it redefines what it is to think. If the stick can sense the ground, is the blind man thinking when he senses it (Kohn 2013)? Rats alter human bodies as they interact sensorially with the minefields, and much of this parallels the ways in which they transform (or transgress?) the relations between species, nations, and enemies to the aspirations of postwar.

The stories in this chapter further explore the untrustworthiness of postwar ecologies and the way war shaped relationships, both nonhuman and human. They also describe how APOPO, the landmine-detection-rat NGO, attempted to push back against this militarism. This led to a marked contrast of the materialities of postwar ecology and the NGO's own publicity materials, which often contrasted with the origin story of the NGO staff.

NGO staff were still former soldiers despite having methods and visuals of mine detection that sought to break from military practice. The NGO entered the field needing to partner with the government's land-mine-detection units, the Cambodian Mine Action Center (CMAC). According to Cambodian law, all NGOs had to have a partnership with a government organization. APOPO's partnership with CMAC entailed an international NGO partnering with a military branch of a foreign government, because CMAC is a branch of the Cambodian military.

The NGO and the national military group, CMAC, evoked the friction between a global industry and its local iterations, beyond conflicts over practice and discourses (Tsing 2005). The deminers at the bottom of the command hierarchy, the grunts on the platoon who did the actual dirty work of demining, often resisted the foreign authority of the NGO. Sometimes the Cambodians deminers regarded the NGO's technologies as having foreign origins—for example, the rats signified "African" and the kennels, made in Cambodia, stood for Cambodia. The APOPO international supervisors from Mozambique, England, and New Zealand all told me that APOPO wanted to make clear that it was not a military presence. It designed its uniforms and promotional materials with this in mind. APOPO was, however, what Cambodians considered a foreign presence. It also became clear that the look of the rats, the look of some NGO workers, and the look of the uniforms affected how the state workers and the NGO workers interacted with each other. These interactions were also influenced by the unseen dangers of the minefields, the bombs. Thus while the minefield as a postwar ecology included the backstories of soldiers, the violence of events in the minefield, and the tensions between people, the materialities and relationships began to push against the constrictions of war worlding. In the setting of the postwar ecology, the land's materialities become imbricated in relationships of power and war. The NGO's presence pushed back against these postwar shapes, even though postwar guided how animals were chosen to be actors on the minefield and how humans understood their place on it.

·   ·   ·   ·   ·

I was unused to how the morning light bathed the baroquely tiled government building that housed the local demining unit—normally we left for the mine-detection training ground before sunrise. I watched a man in a blue military uniform set down a small radio near the central flagpole. The radio began to crackle with some station's music that I did not recognize. Men gathered. A deminer, a man employed by the state to detect and detonate landmines, ushered me to join one of the two parallel lines of men. A commanding officer stood between us.

The officer barked an order and the men clicked their heels together and stood up straight. A song began its crescendo on the radio, and I realized that we were listening to the Cambodian national anthem. The men and I stood at attention as the Cambodian flag got pulled up the pole. The song ended, the officer barked some more orders, and the deminers and I dispersed.

The militarism inherent in the structures and experience of mine action provides us with a common paradox: militaristic disarmament. While disarmament seems as though it is a process of peace, it entails peacekeeping through militarized groups like UN troops or humanitarian aid workers who receive military training as a professionalization requirement. The paradox of militarized disarmament represents two conflicting "regimes of truth." On the one hand, military personnel do not trust peacekeeping activities because of their wartime experiences (Gusterson 1996, 7–10); on the other hand, peacekeeping activities require their practitioners to "buy in" to peacekeeping—necessitating an idealism that soldiers do not normally have. In Cambodia, two groups work at the minefields where rats are employed: the state-employed staff and the NGO-employed staff. We can separate these groups according to contrasting regimes of truth: (1) local state workers who consider themselves to be soldiers in Cambodia and (2) foreign and local humanitarian aid workers moved by global crises to provide mine action. The personnel on the minefield represented two organizations with conflicting ethos and the same goal of disarmament. One can see this conflict most clearly with the introduction of the rat to Cambodian mine action, because the rat accompanied and personified a new demilitarized approach to disarmament. Before the rats, though, militarism was exemplified in Cambodian demining by rituals like the morning salute at the demining unit building.

Shortly after the deminers saluted the flag, staff from an international demining NGO arrived in their van. We drove off to kennels where a new technology was being introduced to Cambodia, the rat. Along with this new mine-detection technique came beings and materialities that announced their implementation through their visualities as something new, foreign, and demilitarized.

We drove forty-five minutes from Siem Reap to Roluos District, a village famous to insiders for being the residence of the renowned architect Vann Molyvann. Like all the villages outside Siem Reap, it was wealthier than the ones I had lived in on the outskirts of the country. Its roads were unpaved, but its market bustled with fat fish flopping in plastic baskets, lush veggies and bright fruits, freshly slaughtered pigs, kitchen appliances, children's toys, and the smells of cooking chilies and fried food. Tourists did not frequent this area, but the residents benefitted from tourist jobs and the increased government and development work that came with the tourist industry. A long, flat, sandy driveway brought us to a pair of government buildings. One was low and flat, and the other was more ostentatious, taller with swooping roof structures and fancier tiles on its front stairs.

"The governor owns that land right there," Hien said, pointing to the acres of flat earth east of the fancier building on the right. "In there is where the government hosts mine action conferences. This is all CMAC," Hien said, moving his hands to encompass the two buildings.

Behind the buildings were the minefields that Liz, the British consultant, and Hien, the platoon leader, had planted. A statue of Buddha contemplated the minefields under a small pagoda. The visual structures on this practice minefield established power relations between the government and the NGO that was importing the rats. CMAC provided the NGO its infrastructure, half of its rat handlers, as well as supervision and oversight. In addition to that was the sight of Buddha, a nod to the religious backing of these powers, which I found out more about when I encountered ghosts on the minefield. Eventually, the Cambodian Mine Action Association (CMAA), a different government organization with multiple ties to CMAC, would oversee APOPO's national certification. This certification was required for the landmine-detection-rat teams to bid for grants to clear minefields. It would take a year before the rats were nationally certified. During that year, the power dynamics and military infrastructures into which APOPO had ventured were rendered clear.

The military state loomed over most minefields, and even the nonhuman animal companions reflected the militarized branches' influence over clearance. The dogs, a common biological technology used by CMAC already, were usually depicted as loyal companions who sat upright with their one handler. When comparing the two landmine-detection animals, the differences in their representations, such as photos and other publicity materials, were striking. These differences were partly due to the difference in organizational ethos, but they were also due to the connotations of rats and dogs in Cambodia and even more widely.

Rats are best known as laboratory animals, cartoons, and pests.[2] In the legend of the Chinese zodiac, the Jade Emperor called for a race that included crossing a river. The rat could not swim, so he convinced the ox to help him traverse it. At the very last moment, he moved onto the ox's nose, putting himself in first place and the ox in second. The rat here is depicted as a clever beast but also kind of *an asshole*. The rat is generally not considered an honorable companion and certainly not one suited for military aid. In fact, the rat is a technology considered by the NGO as a key component of scientific inquiry; for example, I was told that the rats couldn't have Ebola (despite the fears of the Cambodian military director) because they were "lab rats." The rat on the minefield, for the deminers, was a tool that literally modified their untrustworthy sense of sight by letting them "see" what could not be seen.

Military dogs, on the other hand, have been a familiar sight throughout human history. Take, for instance, the Belgian Shepherd Malinois, like Costa, the sleek dog I met in the previous chapter, traditionally used in landmine detection. In contrast to rats, mine-action canines were represented as military technologies rather than scientific technologies. The dichotomy between cute rats and military dogs demonstrates that it is not simply the attributes of the animals but the specific attributes that are emphasized to represent the ethos underpinning mine action. This is apparent in the dog's characteristics and also in the mine-detection techniques used with dogs. In mine action, a dog is related to as a slightly dangerous animal, a creature who trusts only one person. This makes the animal trustworthy only with one person. Recall that when I visited a dog on a landmine-detection platoon, Brahm said, "Be careful! Don't approach the dog until the handler says it's okay!" Moreover, the technique of a mine-detection dog paired to and bonded especially with one handler allows for further militarism by depicting the dog and the human as a striking and imposing pair. In contrast, the rat is on a string and forces the deminers to work in pairs and perform an awkward wide-stepped dance across from each other.

The different approaches to training the two animals is another way to understand this contrast. In their youth, the mine-detection rats become habituated to humans, and then trainers use clicker-training reinforcement so that rats associate the smell of TNT with steps that lead them to a banana. Rat should learn to (a) smell TNT, (b) scratch the soil twice over that TNT, and (c) hear a click. The click becomes an equivalent to the taste of banana for the rat through intermittent positive reinforcement (for more details, see Mahoney et al. 2015). This contrasts with the training of mine-detections dogs. Mine-detection dogs are trained using "alpha" training methods that, while they do not involve negative reinforcement, do involve a handler being the alpha—that is, the "pack leader"—to the dog. These trainings are referred to on websites and pamphlets that feature the landmine-detection dog platoons. The dogs also get trained with only two deminers at a time, and even this pair implies the militaristic ethos of mine action in Cambodia—a dog can bond with only one or two people, but if one deminer dies in the minefield, then with redundancy tactics, another can take over.

Over the course of my years of fieldwork, however, I have noticed a change in the representations of the dogs that seems to be inspired by the successes of APOPO's visitor center and promotional materials. Recent Malinois landmine-detection promotional materials—featuring cute puppies and wagging tongues—seem to be trying to play off the success of the rats. APOPO's "HeroRATs" turn the traditional military nature of landmine operations on its head. APOPO takes every opportunity to capitalize on the playfulness of the rodents, holding HeroRAT "adoptions" in exchange for monthly or annual donations and sending adopters an email containing a "Top Secret File." Demilitarization, the APOPO staff told me, is key to the success of landmine detection. They explained that military procedures, with their dependence on hierarchy and redundant chains of command, slow the detection process, putting more lives at risk. APOPO promotes such demilitarization not only in the design of its detection techniques but also in the visuals it presents. Playfulness entered a context of militarism and suspicion that came from the ways in which people (dis)trusted each other in a postwar ecology such as a minefield.

.   .   .   .   .

Moch's gentleness belied a history of a home disrupted by explosives. Just beyond Moch's childhood home, a famous temple, Preah Vihear, straddled a national border militarized since the late twentieth century. The temple was a complex of buildings from the ninth century, from when Cambodia was an expansive empire. On a visit to the temple, we took a military jeep up a steep slope. As we ascended, the food and drinks we brought with us slid backward past our feet. The refreshments were not for us, but a customary kindness for the soldiers, like Moch's father, who guarded the temple. The conflict that led to the border closure had long since ended, but its traces and those of earlier conflicts remained embedded in the landscape—a deadly palimpsest.

At Preah Vihear, a long stone pathway took us up the mountain. Trail lines guided my eyes to the silhouettes of temple tops. The plateau comprised a complex of stone buildings sculpted with giant carvings of flowers and gates etched with a story of gods. Splotches of white and chartreuse lichen embossed stone details. The temple had clean, symmetrical lines with rectangular windows and halls. Wind and cloud mists flowed through its entrances, but our movements were limited. Beyond the path, Moch told me, there were landmines.[3]

Landmines, which have a lifespan of one hundred years, persist longer than the conflicts that motivated their installation. They outlive their original uses; they take (on) different lives. Today, landmines at Preah Vihear prevent Thai access to the temple complex and have become part of a border dispute that developed after they were laid, according to the deminers I've spoken with. This series of events and the resultant layers and possibilities of unexploded ordnance remind us that conflicts do not start and end, but rather nest and beget each other. They hold themselves. Unlike concrete walls, landmines cannot be torn down but must be detected and detonated, piece by piece. Decontamination is methodical, tedious, and tactile. Inhabiting a contaminated place like Preah Vihear meant that borders were physically reinforced between nations but transformed people, deconstructing divisions between life and death.

Moch told me the stories her father had told her. Before she left to be trained with the landmine-detection rats, she had been raised on his stories about being a soldier in Preah Vihear. "I want to help my country," she told me when I asked her why she became a deminer. Her father's stories that she relayed to me were not only about comradery between fellow countrymen but even between enemies separated by national borders.

The Thai soldiers and the Khmer soldiers had a designated area where they ate meals together. At Preah Vihear, I asked a local soldier about it and he pointed to the area. Moch explained that her father's fellows-in-arms and the Thai soldiers would celebrate holidays with each other, too— Khmer New Year and Pchum Ben. While the two sides were stuck on post because they were supposed to keep the area safe from each other, they had as a consequence become neighbors and family. This did not mean they would not shoot each other if their officers gave the order. And it did not mean that they are not proud of being Khmer (or Thai, I suppose).[4]

Vulnerabilities, even though they were enemies, were shared due to the landmines that haunted them both. These landmines are unmapped and unknown to either side. Even though the soldiers sometimes reused the bombs against each other, they also sometimes warned each other about them. One Cambodian soldier told a Thai soldier to stay away from the west side where bombs lay from the 1980s while another collected the bombs and used them to attack another.

When I visited Preah Vihear with Moch, after we walked up the long stone pathway, I noticed a faded sign lying on the ground. It had the familiar skull and crossbones on the smooth red and purple black stony surface.

"Oh! There aren't mines *right* here, are there?" I exclaimed. It looked far closer to the path than I had expected and its presence surprised me despite Moch's warning. I felt my body tighten. The very threat of the landmine held my body. I felt surrounded and immobilized. It did not yet touch me but its force was pushing me away from it. Landmines touch you in that they push you around, move you, even when you don't touch them.

A soldier who had accompanied us on our tour turned to me. "There are," he said, "Someone was just hit by one on the path."

"When?"

"Just last month." A tourist last week, the soldier told us, had even asked if he could go down a path to the west, and the soldier joked with us, saying that he told him, "You can, but that was where the bomb exploded."

We walked a different path that day.

The bombs at Preah Vihear both reinforced the boundaries between nation-states and allowed them to be transgressed. The landmines kept Thai and Khmer soldiers behind enemy lines but they also configured their spaces—here was a safe space, a cleared space where the enemy soldiers would pray for the dead together during Pchum Ben. They told each other where these explosives were while simultaneously removing and changing the border in a shifting line[5] to attack each other. The bombs enforced divisions and suspicions.

I am not the only person in Cambodia who has felt frozen by the threat of a bomb. Many villagers and deminers I spoke with described similar sensations in their bodies. "My heart just got cold," gasped a woman I was with when a bomb interrupted our path.

This disruption is not only because landmines are a danger to a person's body but also because they "lie in wait" in places where they are unexpected (Kim 2016). These landmines now perpetuate the violence of war in times as well as spaces of peace. To destabilize peacefulness in space and time is also to destabilize fixed ideas about being human. Now, when a villager stops for a drink or visits a holy site like Preah Vihear, he can lose a leg. The landmines, for deminers, speak to the perpetual tendency of humans to be violent. This is something that often corresponds to their personal histories in Cambodia, as many of them are former soldiers who fought against each other. And yet, these technologies have now transformed their relationships from enemy to coworker, forming postwar relations.

.   .   .   .   .

Hien tapped me on the shoulder, and when I turned, I was face to face with a stuffed rat.

"We have no rats, yet, so we practice with a stuffy." He laughed and motioned for me to follow him outside.

The pits below Buddha's gaze were outlined in twine connected to a series of metal posts. More posts seemed randomly aligned along the borders of the pits. Some of them had black numbers and some red.

"The numbers," Hien explained, "show where something has been buried. So there is a number five. That means five feet into the pit, there is either a piece of metal or a bomb buried. If the number is black, it is a piece of metal. We want to test false positives. If the number is red, it is a landmine."

I looked at the red number. I thought about the explosive just under the earth.

"It's deactivated," Hien said, perhaps seeing the flash of fear on my face. "So don't worry. This is just a practice minefield."

As a field researcher in Cambodia among mine-action groups, I had heard rumors about deactivated landmines that had exploded because of some mistake. One person had even told me a frightening story. The person had ordered a bunch of landmines from the government's storage facility for training. Hundreds of landmines had been transported six hours on the bumpy, unpaved roads. Then, once they arrived at the training field, the truck was emptied and all the explosives were put on a table. It wasn't until an experienced deminer saw the landmines and, shaking his head, said, "None of these are deactivated!" While that story miraculously ended with no accidental explosions, I still felt fear seeing the red numbers on the practice minefield. I had worked in minefields long enough to know not to trust the ground I walked on.[6] The numbers rendered the invisible visible to certify whether or not the rats were able to correctly detect landmines.

Hien demonstrated a technique of putting the rat on a restructured fishing pole and a loop. This would be a technique that did not work for the rats when they came to Cambodia, and it would be abandoned. The rat, in this case, was supposed to be guided in wide semicircles radiating from a single human deminer. This technique basically replaced sweeping a metal detector with sweeping a rat at the end of a stick.

Hien's demonstration of the sweeping technique, though, was months ahead of the rats' arrival, before anyone knew the rats would not cooperate as planned. At this point in time, the minefield was a stage on which the NGO staff prepared for the main show.

Such a staging was important to establish the NGO in Cambodia. The practice minefield and Hien's uniform and equipment were to demonstrate to CMAC, APOPO's government partner, that the NGO had the funding and capacity to bring mine-clearance operations to Cambodia. The relationship with CMAC was uneasy, though, since the institutions differed in their approach to demining. As noted earlier, CMAC was embedded in military structures in Cambodia, whereas APOPO considered itself to practice scientific, technologically innovative demining.[7] In addition to its search for certification, APOPO had to prove its legitimacy as a demining group, despite its lack of militarized style.

For Hien's demonstration, there were no rats, only the settings for the rats and the stuffy. The stuffy obeyed the movement of the pole, and we all laughed at the children's toy as it hovered above the ground, bending the groomed grass briefly.

Liz appeared outside the door of the larger building. "Oy!" she shouted on the steps, "Want to come see inside? Kind of interesting."

Hien motioned for us to go, and Chann, who was the manager above Hien, and I left to go to the other, more ceremonial office building. The steps led to a green tiled foyer with a massive varnished wood table and chairs. The heavy furniture stood to the side of some tiled stairs, but Liz directed me to a glass cabinet on the left wall: "These are just a bunch of explosives they've found."

The cabinet was floor to ceiling and held shelves and shelves of bombs. Some of them were antitank mines, big rectangular things. Some were grenades, tiny metal pineapples. There were two giant rocket missiles with rusted red metal, pointing down at the ground. But dozens of the bombs on display were antipersonnel mines—different sizes and materials, some metal, though most were plastic discs. Most were the circumference and thickness of a smoke alarm (figure 4). They were labeled in both Khmer and English with rectangular black signs saying what kind of weapon they were and where they had been found. "It's kind of like a museum," Liz said. "When they have official conferences here like from Geneva, the visitors come and see this exhibition. But it's also good for teaching and training deminers."

Figure 4. Museum display of bombs outside Demining Headquarters, Siem Reap. Photo by author, 2015.

Here the landmines, things that normally went unseen, were on display both as proof of expertise and pedagogical tools. The bombs were a visual array to demonstrate to international visitors the legitimacy of the problem of landmines in the country as well as the effectiveness of the major player in the national landmine-clearance industry, CMAC. The officials from Geneva whom Liz referred to were grants advisors and program officers from the Geneva International Centre for Humanitarian Demining (GICHD). The GICHD organized conferences and distributed funding from international donors to various mine-affected countries. CMAC regularly applied for grants and courses through GICHD, and the glass cases were important visual reminders to international visitors of the need for those grants and courses in Cambodia. The minefield where the rats would be trained were owned by the Cambodian government and used as a performance to network with international funding bodies.

The display rendered the invisible visible—these explosives behind glass were still untouchable, but they portrayed the extent of Cambodia's explosive-remnants-of-war (ERW) problems. They were organized chronologically through the various wars that had left such remnants, from the American bomb dropping during the Vietnam War through the 1985 K5 landmine installation.

We drove back to Siem Reap. Liz introduced me to other officials at the Siem Reap offices. She gave me an office with Hien and Chann, where I could work at one of those heavily varnished wooden tables.

Even though the rats hadn't arrived, I followed the schedule of the deminers who worked in the office revising the Standard Operating Protocols (SOP) and planning the training. The SOP was what the deminers used to keep safe. Paul, a New Zealand demining consultant, Hien, and Chann were all revising it for APOPO. Paul explained to me, "Normally, these SOPs are written by military guys. They hold onto a lot of military practices like redundancy, meaning that every task must be taught to multiple people so that one person can take over if another dies. But this is not wartime! The deminers aren't in any real danger; they don't need built-in redundancy. That just costs extra to train people extra."

Early mornings, I rode a brakeless bike through the traffic of tuktuks and motos along the Siem Reap River, which transected the downtown area. It was fairly safe to ride such a bike in this area as the tuktuks and motos drove slowly, calling after tourists for rides to the temples. The bike ride to the office took about forty minutes. After I arrived, I read NGO documents in the office, a white tiled room on the second floor of the CMAC government building.

I was one of only five people not wearing government military uniforms in the building. CMAC had agreed to work with the NGO APOPO. Although Liz didn't wear a uniform, Paul, hired directly from the New Zealand military to revise the military strategies of landmine detection, and Chann and Hien wore uniforms issued by APOPO. The uniform, while not as militaristic as CMAC uniforms, with gold fringe on their shoulders and culottes, still had a cut that indicated army personnel. Their plain cotton T-shirts and cargo pants strayed from the military visuals, though, and foreshadowed a new choreography that would come with the rats.

"We set up the minefield so it mimicked the layout of contaminated pits—200-meter-by-200-meter pits. Between them are pathways that are already cleared," Hien explained in English because Liz was there. Liz nodded. "The deminers, even though they are only practicing, can only walk on the cleared areas."

We walked behind the building. The handlers had not yet been hired, but APOPO was beginning its implementation of the new rat detection technique. Hien and Chann, with decades of demining experience, had been hired by APOPO to travel to Mozambique and Tanzania to train with the rats there. Liz, who had worked in mine action in Southeast Asia since the 1980s, had been hired to oversee the implementation of the rats in Southeast Asia. She had tried in Laos and Thailand, but it was in Cambodia where the NGO had had the most success. She saluted us, two fingers on her forehead. "I'm going to say hi to the CMAC officers over there. Come by after your tour."

Hien and Chann guided me into the building on the left. It was painted pale pink and had a short staircase to its tiled porch for the rainy season floods. There were three rooms, and the first room we went into was set up for the rats.

Large cages, each big enough for a fifty-pound dog, stood on wooden platforms. The two deminers took me to each cage to explain the items in them. A drip water bottle hung by wire next to each door. A terra cotta pot with a flat side would be the bed for the rat to sleep in and feel safe in, because "they like the dark and snuggle in there." Sawdust at the bottom of the cage would help them make their nests and absorb their urine and feces. A triangle of wood stood next to each terra cotta pot. The walls of the room were yellow and on the glassless windows there were red fleece blankets to block the sun, which leaked through the blankets, making the light rosy.

"Everything was specially made by Cambodian villagers—the cages, the pots, the toy," Chann said. I asked what the wood triangle was for and he took one out of the cages. "This is a toy—the rats need to chew. We had to ask a local carpenter to make it specially."

Hien insisted on locally made craftsmanship. It reminded me of the ways in which he and Chann seemed critical of the food and weather of Tanzania. Their month-long training was the first time either had been out of the country. Both shrugged when I had asked them if they had liked their experiences. They had missed their families, language, and food too much to enjoy it. This insistence on using the craftsmanship of Cambodian workers seemed to go hand in hand with this dislike of their visits abroad, but it also went hand in hand with a kind of pride that countered the "Khmer killing Khmer" history.

The resistance to foreignness, though, was not only on a staff level and alluded to enmities. APOPO's attempt to gain buy-in from its required government partners met with considerable resistance, too. The first arrival from outside Southeast Asia for the landmine-detection-rat implementation were the Tanzanian consultants, Mohammed and Daudi, and the Mozambican consultant, Clarence. An American animal behavioralist, Brad, who lived in Mozambique, was also coming. This took a bit of bureaucratic finagling, and it became clear that the extra hurdles were about colorism and racism. It was far easier for Brad to get to Cambodia than for any of the three native Black Africans. Paul, the military consultant from New Zealand, was more open about discussing this than Liz. Liz only said, "I don't want to say the *r* word, but . . . " She drew on a cigarette instead of directly clarifying that the *r* word was *racist*.

"They only got through customs when I showed up, all smiles, telling them what a good job they're doing and what a beautiful country Cambodia is and how excited we are to show our visitors this beautiful country," she said. The customs officials had told her that because Cambodia had so many drug problems tied to Nigerians traveling into Cambodia that they had to stop everyone from Africa. Even with the official CMAC positions and letters of offer and money, they couldn't get a business visa. Instead, Daudi, Mohammed, and Clarence all had to get a one-month tourist visa. This would be annoying because every month the organization would have to get the three of them out of the country and then have them re-enter to renew their tourist visas—what was commonly called a "visa run" by the longer-term visitors.

Liz worried about sending them to Vietnam, thinking that they'd have the same problems. She said they'd probably have to send them to Hong Kong for visa runs since the system was more standard. Liz's success as a mitigating factor implied racist sentiments, though, because her presence and whiteness afforded the Black Africans passage through customs.

Paul was more up front about the prejudice against Africans that slowed down the import of the rats, but he also was clear that this prejudice was part of systemic issues in Cambodia's NGO world. He told me stories of witnessing European NGO workers hitting Cambodian staff. Racial prejudice took on a colonial, systemic pattern, where Cambodians expressed racism against Africans in a kind of hierarchy that could be imagined as inherited. The look of importing an "African" technology with Black supervisors resulted in uncomfortable barriers for APOPO's implementation process. For instance, the stories about white European commanders slapping Southeast Asian staff seemed linked to the legacy of colonial supremacy that manifested as colorism among Cambodians when they expressed frustration with the dark-skinned Tanzanian consultants and acceptance of me, a white woman, and of Liz and Paul, who were both white. While there are certainly histories of colorism that originate in Asia, scholars have pointed out that colorism in Asia today is inextricable from the Western colonial legacy of racism, especially in the postcolonial context of South and Southeast Asia (Jayawardene 2016). The colorist hierarchy itself reflected not only colonial values but also seemed connected to its militaristic hierarchy. Militarism depends on enmity in its structures and ethos.

Racism itself is a colonial legacy that operates under the logic of white supremacy. That is, the racial hierarchy Paul refers to originates in the ideas inherited from French colonial norms (Fanon 1963; Edwards 2007). At the same time, colorism has historical roots across Asia from before colonial impositions (Bettache 2020). Colorism also has an uneasy history in Cambodia since, during the Khmer Rouge regime, colorism was reversed in its valuing darker skin over lighter skin, since lighter skin was associated with foreign influence as well as a product of elitism that opposed the Maoist communist valuing of the farmer. Today, it is not uncommon to nickname a darker-skinned person Mao, which translates to "Black," but then again, Yeay Mao, who appears in a later chapter, is a powerful grandmother spirit warrior who is quite high in a hierarchy of power. Colorism becomes complicated by these histories, but it was still clear that the foreignness and Blackness of "African rats" posed a problem for their import as well as creating interpersonal staff issues with the African consultants who came to Cambodia to train the deminers.

·   ·   ·   ·   ·

Importing the rats for the first time took over six months. Liz and Paul had been at the forefront of this process, negotiating the memorandum of understanding with the director general (DG) of CMAC in the Cambodian army. The general had resisted the import. Paul was driving me to see the kennels again, and he said suddenly: "We were about to give up because we didn't know why we were here. . . . The budget's been cut—partly because of donor fatigue" but also because " the donors aren't stupid," and they wanted to see their money used more effectively.

According to Paul, the rats were so much more efficient that the organization could even lose money. This, he speculated, was part of the reason that the DG resisted the import. Though they were partnered with CMAC, Paul complained that the DG wanted more control over everything.

"The last excuse they had for delaying the rats import was Ebola!" While driving, he lifted his arm in frustration, "First," he said, "they're rats from a lab, and second, it's kilometers away from Ebola! There's no way they carry Ebola. And of course the general made racist comments about bringing the trainers in from Africa to teach Khmer people how to work with rats!"

"But," he said, seeming to correct himself in front of me, "it was just one comment, and the African guys got their letters, no problem. The reason there are suddenly no delays is because they saw that we were threatening to go and pull out some money. All of sudden the bureaucracy moves more smoothly!"

It was in this conversation that Paul also noted that Brad, the white animal behaviorist coming from Tanzania would be able to get his visa much more easily than the Black African consultants. Notably, Brad was referred to as *barang,* the Khmer word for "white."[8] This was all part of the postwar ecology, a relationship framework built on nationalism and hierarchy that determined who and what was acceptable.

Since the problem with the import seemed to be about their foreignness, I asked APOPO staff why they were using these rats and not local Cambodian rats. "Well, there were only three options in Cambodia—two arboreal species that cannot smell as well and one species called the Bamboo Rat. The Bamboo Rat is just not as cute. People like our rats" (see figures 5 and 6). Tree-dwelling rats also presented a problem because they are less scent-oriented than ground-dwelling rats.

Figure 5. Bamboo rat, Khaosok Zoo. https://www.khaosok.com /national-park/bamboo-rat.

Figure 6. A landmine-detection rat eats out of a deminer's hand. Screenshot from film by author, 2015.

After researching images of Bamboo Rats, I realized just how impor-tant the cuteness factor was for APOPO's rats. I agreed with Brad that the Bamboo Rats were difficult to "cutify." Behavioral scientists have con-nected aesthetics of "cuteness" with neonatal attributes such as big eyes and foreheads (Grauerholz 2007). Though the universalism of this claim may be debatable, the APOPO rats were indeed drawn with such features and full cheeks, and they are often photographed as cuddling with their handlers. The Bamboo Rat, in contrast, has beady eyes and large teeth (figure 5). I have heard this species of rat described as "cute" by academic colleagues but not by anyone associated with landmine detection versus the cute landmine detection rats (figure 6).

APOPO's focus on appearance was not merely being superficial—these visuals were key to APOPO's marketing campaigns in and beyond Cambodia. Thus, from an institutional perspective, it was worth it to seem foreign to their Cambodian partners with an African rat because APOPO attracted donors through the rat's capacity to render demining an innovative and appealing process. These attributes of innovation and cuteness contrasted with the general military aesthetic of the landmine-detection industry worldwide and particularly in Cambodia. An emergent ethos of demilitarized disarmament became possible through the attributes, both historical and physical, of the mine-detection rat. By *ethos*, I mean what matters for the organizations, such as the militaristic values that enable Cambodia's mine-action organizations. The mine-detection rat is unique among the materiality and networks of demining: it does not have a military history, military connotations, or military experiences. In this uniqueness, it comes up against an ethos of militarism in mine action, which is not limited to Cambodia but is particularly entrenched there.

Disarmament depends on militarism, requiring military skillsets (explosive knowledge), military materials (landmines, demining vehicles), and military structures (commanding officers, platoons). These military skillsets, materials, and structures drive mine action's organizations, meaning that postwar becomes even more clearly an aspiration rather than a contemporary reality when the infrastructures are entrenched in war.

It is through the rat itself that demilitarization becomes possible within the entrenched militarism of mine action. The rats are not only innovative as a technology but also enact emergent ideas about mine detection that contrast with militarism. APOPO deliberately attempted to contest the militaristic ethos of mine detection in Cambodia. Demilitarization as an ethos was enhanced when the NGO portrayed mine-detection rats as cute and scientific, but demilitarization was also amplified in unexpected ways by the attributes of the rat and the role the rat played in the new mine-detection technique.

APOPO portrayed the Gambian Pouched Rats as "HeroRATS" that disarm countries. They were featured on social media being held by and crawling on the deminers, domesticating rat and humanizing deminer for international and local audiences. Landmine-detection rats entered a scene where their visual presence helped to demilitarize landmine detection. The mine-detection rats are "cute," and this cuteness worked to demilitarize disarmament both in publicity materials and in the technique of mine detection with rats. Rats who must work closely with humans and, in fact, save the lives of these humans become cutified—they show up on APOPO's website as adorable little animals to be dressed up for Christmas in knitted hats, as cartoons wearing capes, and animals who do anything for food.

Although both biological anthropologists and zoologists have suggested that cute animals have a universal aesthetic in that they have infantile features, wide eyes, and symmetric faces (Sherman, Haidt, and Coan 2009; Estren 2012; Nittono et al. 2012), other human-animal researchers have pointed out that "cutification" is a process. Organizations that market meat, for example, transform animals to be "cuter" through anthropomorphism, emphasizing neoteny, or making them cartoonish (Grauerholz 2007). This contrasts with how animals are "de-animalized" for the workers who must slaughter them (Hamilton and McCabe 2016).

It is not that APOPO's rats are ugly, but they are not typical of cute animals like baby pigs, dogs, or cats. APOPO's rats come from Tanzania. They are of the phylogenic order Rodentia, the genus *cricetomys*, and known by their common name of either African Giant Pouched Rat or Gambian Pouched Rat. They weigh, on average, 1–2 kilos (2–5 pounds) and they can grow up to 0.9 meter (or 3 feet) long, although their two-toned tails make up half that length. They look like typical urban rats except bigger— they have everything you would expect a rat to have: circular folded translucent ears, beady black eyes, tiny handlike paws, and pointy noses that quiver constantly since scent is their most dominant sense. Because APOPO's rats are ground rats, their sense of smell has evolutionarily developed to be better than that of arboreal rats. As rodents, they are both smart and trainable, but because they are giant rats, they are large enough to be controllable. Unlike other rats, such as the commonly seen Norway rat, Gambian pouched rats have big cheeks like squirrels. The rats sometimes stuff these cheeks full of food for safekeeping, giving them fat faces. This helped APOPO emphasize their cuteness.

APOPO's cutification worked to "animalize" (or even humanize) the
rats in featured publicity materials, while it also humanized deminers.
The rats are front and center in most publicity photographs and social
media, though they are only named as individuals when they are shown
without humans. The person in the photograph allows the rat a human-
like identity. The rat becomes an animal who cuddles and nuzzles against
her humans who have serious jobs. In turn, the rats humanize the people
of mine action so that they become transformed by the animals associated
with them—deminers are not only soldiers but friendly-faced men and
women with a rat on their shoulders. The rats de-emphasize these humans'
military pasts for international audiences. Rats are shown as animals who
will do anything for anyone who gives them food. The NGO uses these
characteristics of the rats to its advantage and in pursuit of its demilitarization visuals—rats are adorable, insatiable creatures. The materials play
up the stereotype of a rat as an animal that eats all the food humans throw
away to become a baby-like being who needs to be fed all the time. All this,
however, contrasts with the normal untrustworthiness of postwar ecologies, where they did not fit in.

Several months after Paul and Liz's troubles with importing the rats,
the problems started on the Tanzanian side. Paul and Liz had both
described the tedious work their coworker was doing in Tanzania, every
morning going to the customs office to ask for the papers to be signed for
the rat export.

.   .   .   .   .

The sun glared on the Thai side of the border, more brilliant there, where
the street was paved, unlike the Khmer dirt road in Poipet. We peered
across it, some of us shorter than the others, standing on tiptoes. We were
watching out for Chann, who had managed the rat's importation across
the border into Cambodia. The APOPO staff and I had our cameras ready.
APOPO wanted us to document the rats when they arrived in Cambodia
for the first time, but the border was crowded with people, vans, and cars.
We were looking for a man who would be pulling a cart behind him—a
cart like the ones used to sell spicy snails on the street. He was a Khmer
villager whom Chann had hired to transport the rats across the border in

temporary cages made of jerry cans. Import laws meant that they could not bring them into Cambodia in the air-conditioned van.

Chann stayed with them all night; they had spent the night in an air-conditioned room. He described them to us with glee in English to Liz and to Brad. We were all speaking English because Liz and Brad were both monolingual anglophones. "I couldn't sleep," Chann said. "They squeaked all night—sometimes fighting with each other through their cages. One bit me through the cage."

"Because they don't sleep at night—they're nocturnal. I think they're also on high alert because of all the different smells and sounds," Brad said. Of all of us, he was the most worried—they had been so sleepy when they disembarked from the plane, he saw on the video chat. He was worried it was too hot for them, that they had been through so much trauma.

Hien, who had already met the rats in Tanzania, came running to us. "Hey! You missed us," he said. Behind him was a man who looked like a farmer, his brown clothing hanging off thin limbs. He had a wide-brimmed straw hat to keep the sun off his face. And behind him was the cart.

We all rushed to see the rats. We peeked under the canvas that covered the cages, but Brad waved us to the van where we could cool them off. The three men began transporting each jerry can cage one by one. The rats were stirring, brown creatures surrounded by the bright yellow jerry plastic that was cut to make a box. Red wire covered the jerry cans' entrance so they couldn't jump out. They stuck their noses through the red wires, sniffing. Their little handlike paws clutched at the wires.

"Welcome to your new home," Chann said, "very different from Tanzania. Not the same place."

Seeing the rats made all the difference to the team. The staging was no longer a performance but an intervention that was about to be implemented. Finally, the rat handlers, the lower-ranking staff, could be hired. This is when the candidates had their trial by touch to assess their comfort level with the rats. This was when Daudi, the Tanzanian consultant, and Hien, the Cambodian platoon leader, had to work together to plan how to manage the training. The first sight of the rats was the beginning of the collaboration between international and militaristic divisions. It also brought former enemies together on the same team. Until that point the divisions had been NGO versus state, domestic versus foreign, after the arrival of the rats, the rest of the rat handlers were hired, and this brought a new division onto the minefield, that of former enemy combatants. These layers of division were now a mesh of NGO versus state, domestic versus foreign, and enemy versus enemy.

.   .   .   .   .

Brad wanted the rats to have two weeks to adjust before the training began. This was pushing it for the platoon, who were very antsy to get out of the office. Liz smoked a lot outside, pacing and talking on the phone. They would hold the interviews that week, and the training would begin immediately. She had to act as a go-between, fielding demands from both Tanzania and Cambodia.

APOPO went through more staff transitions. Paul was leaving; he wanted to return to New Zealand. The NGO hired a Cambodian man to take care of the rats. Tokla was a vet who spoke some English and could communicate well with Brad. He had mostly taken care of pigs in a small factory farm in a city between Phnom Penh and Siem Reap. His family still lived there, but for this job he moved to Siem Reap. I met him while visiting the rats in their kennels.

Tokla was a short man with carefully gelled hair and wide-spaced teeth. He had a gentle way about him and matched Daudi's attitude toward the creatures. Hien had driven Daudi, Brad, and me to meet the rats, and Tokla was puttering around in the kennels when we arrived.

The room was kept cool by two white air conditioners on the wall, and Tokla had sealed the windows with thick blankets. The sun shone through them, but the shade helped a little to darken the room with rosy light, as if it were sunset. I was so excited to see the animals in their homes that I squealed a little and Tokla laughed at me. "Yes, they are very cute," he said.

I walked with him to see each one of the rats. Most were huddled in their terracotta pots. Brad, the animal behavioralist, walked alongside us and we all crouched a little to peer into the crates. Each time we did, the rat would peek his nose out of the pot, twitching his nose. On every crate was a tag with the name of the rat and the names of its parents. The NGO kept track of how well each family line did, and the rats imported were from the best detective lines, meaning that they had attributes desirable for the landmine-detection work. These were the rats who had had the most success remembering their training (usually, some rats that are trained need "refresher courses") and the best socialization with humans.

I couldn't help but imagine the people behind the names, though. The names were often Christian-sounding, like Mercy, or they were English names like Janthina, and some were those of landmine-detection-organization researchers, like the Norwegian name Håvard, which I knew must be a reference to Håvard Bach, who had written papers on biological technologies like the rats.

I had lots of questions for Tokla and Brad about the rats. "Do they get along with each other?"

"Well, normally they wouldn't, unless they were related, but the rats are kept in adjacent cages with each other from a very young age. So they get used to each other."

Landmines, even though they may or may not explode, make not only the ground beneath your feet untrustworthy but also your friends, your colleagues, your neighbors, and yourself. These relations are suddenly revealed as unsteady, and they illuminate what Talal Asad (2007) and Stanley Cavell (1988) mark as humanity's existential crisis: our kin, our fellows, our very own selves cannot be secure or constant. In the postwar ecology with this new actor and new technology, the landmine-detection rat, relations began to change—from enemy to beloved but also back again (repeat).

4 A Murder

It is not uncommon to find bodies in unmarked graves from Cambodia's civil wars, especially the Khmer Rouge, even in the twenty-first century. Soldiers executed people en masse, and corpses were dumped in pits and left disinterred. These deaths are classified as "bad deaths," and most of the bodies have been unclaimed by families, since they were separated during the Khmer Rouge era and afterward (Kleinman 1995; Han and Das 2015). The ghosts of these former wars haunt the landscapes (Kwon 2008; Zani 2019).

In 2010, before I conducted fieldwork on deminers, I researched villagers and former soldiers of Battambang province at the minefields near the border of Thailand. On the weekends, I would stay in Battambang, a former colonial city on the banks of the sandy-colored river of the Tonlé Sap. During the 1950s, the place had been designed as a vacation spot for rich French people and, for me, its buildings evoked Hemingway's stays in tropical resorts, with mold painting the sides of its stucco walls like an incongruous takeover. Small business owners and families had taken up residence in these buildings, sometimes painting them and sometimes leaving them white, resulting in a mosaic of colors along the river walk. At the time, the city was rarely frequented by foreigners, but its proximity to

Siem Reap meant that its tourist industry was beginning to grow. I was friends with a few local Australian entrepreneurs who were trying to encourage growth there, and because of my Khmer language skills, I was invited to beta test one of the tourist attractions: a bike ride.

Local youth workers at NGOs where I volunteered on the weekends had told me that Battambang had been renowned for its French architecture and that the people used to ride their bikes everywhere. A young man, Seyla, had been hired by the Australian entrepreneurs to guide me to various attractions: from a family that demonstrated how to make rice wraps, to a bat cave, to another family that demonstrated how to make rice wine. One of the last stops was a local killing field.

These killing fields are places where massacred bodies were dumped. Outside Battambang city center, Seyla and I rode our bikes down a sandy narrow path under trees. As the trees thinned out, we arrived at a building where a series of orange robes hung from the railing.

"Monk laundry," Seyla said, his hair flopping as he laughed. Seyla was close to adolescence, skin mottled with acne scars and shiny hair that swooped over his forehead. He flashed his teeth when I laughed at the monk's laundry, a beautiful sight of bright, uniform fabrics. But then Seyla sobered as we continued on foot to a building adjacent to the monk's pagoda.

The trees had been completely cleared, and in the center of the clearing was a small concrete structure, painted white, with steps leading up to gilt trimmed walls with several barred windows. The slanted roof was decorated with gold filigree and deities. Nagas (snake deities) lined its peaks, and garudas (bird deities) spread their wings beneath its pagoda-styled roof peaks like turrets.

"We are entering a sacred place," he said. "There are many spirits here."

He gestured to the building and invited me to approach after I covered my shoulders with a krama, a scarf. I always carried one with me in Cambodia since, as a woman, I needed to cover my body in certain places. I was surprised as I came face to face with a pile of skulls. The place's name on a wooden sign was in both Khmer and English: The Well of Shadows.

When we approached the steps, I stopped to look at them. On each step's riser were bas-relief depictions of the torture and killing that had occurred during the Khmer Rouge. The first was of soldiers taking people's bicycles, followed by soldiers smashing infants' skulls on trees, people being tortured, starved, and forced to marry strangers. After walking up the steps, I found myself looking into the eye sockets of skulls, lined carefully on shelves behind glass windows and metal bars.

The structure was a tomb reconfigured for tourists and visitors. Normally, a tomb in Cambodia would be a tall structure with gold-topped roofs like this one, which held the ashes of a body. It would not have windows. This tomb put the dead on display. The structure was simultaneously evidence of massacres and a memorial to the lives of the massacred. Ashley Thompson (2013) has discussed these contrasting ideas of memorialization and evidence as values that come from survivors' emotional responses to violence—an archive can serve not only to bear witness but also to provide evidence against crime. The archive has its own extrapolations of voyeurism, which come from ideas of justice where perpetrators have evidence stacked against them. As a result, killing fields have been transformed into this uneasy compromise: bodies blessed by a monk and laid to rest in a glass enclosure.

Seyla explained who the skulls had belonged to: "The people buried here are in this building because they used to wander the fields after they died. They were killed during the Khmer Rouge and then their bodies were left here. After the war, it would be very unlucky to come here. People would get sick because the ghosts haunted the field. You could hear them screaming at night. The villagers asked the monks to do something, and so they blessed the ground and built this pagoda for the bodies to rest."

There were traces of these ghosts everywhere. I would hear similar stories from neighboring villagers, too. They would casually explain that the disinterred bodies from the Khmer Rouge had been mounted behind glass as a way to let the spirits rest. One friend, ten years after Seyla and I had ridden our bikes, described the process of silencing the ghosts in almost the exact same words as Seyla had: "The ghosts had been screaming every night. So, the monks blessed the field. Then, the bodies were placed in a mausoleum with glass walls. Now the ghosts are at peace."

Ghosts and spirits often bifurcated in scholarly taxonomies of the spiritual world in Southeast Asia (Tambiah 1975; Eisenbruch 1992). But in people's descriptions of their relations with ghosts and spirits, the two generally overlapped. In fact, they showed up as diffuse and mutable in ways that I understood only after hearing multiple stories and visiting multiple sites imbued with their powers.[1] I came to track the ways in which spirits/ghosts were understood through extrasensory encounters. That is, the ghostly realms as they manifested during fieldwork were sensorial encounters that presented mysteries to the human beings who encountered them. These sensorial encounters were characterized by altered materialities that did not follow rules of more easily understood sensorial encounters. Ghosts or spirits were known to set off bombs unexpectedly. They made their desires and expectations known through altering materialities via murder, screams, unusual apparitions, the mysterious protections they offered through drawings on human skin, and explosions. They were an important part of relations on the minefield, and they were not separate from its materialities or even from the infrastructure of power in the organization. As I will show, the ghosts were part of the hierarchy of power found in the government and in the wars past. They presented ways of understanding entanglements with beings that were potentially violent, paralleling the entanglements of the deminers with their enemies and of the deminers with the bombs laid below ground.

The spirits, once unsettled in the killing fields, interrupting the sleep of the living because they could not partake, were now laid to rest. And their rest was inextricable from the memorialization of the Khmer Rouge genocide. Not all the ghost stories I heard entwined ghosts with the Khmer Rouge and with ways of honoring the war dead, but wars and violence were often parts of ghost stories. These stories were told often, except on the minefield with the rats.

Like the secret backstories of former enmities between coworkers, ghosts were not often talked about on the minefield. The deminers would talk about ghosts or spirits only when I asked about them. I had an interest in *yantra* tattoos, because some of the older veterans I interviewed who later retired from work in minefields wore this shamanic ink that they claimed kept them safe in the minefield (DeAngelo 2019). These tattoos were drawings of spirits who could enter a person's body through the skin. Once they entered, the spirits would protect the bearer from violence or imbue the person with specific powers, such as the deminer I met who had a bird tattoo that would make his voice seductive to women. While the deminers told me stories about how these tattoos worked, none of the members of the rat deminer platoon claimed to have any tattoos except for Meas, who had a tattoo of their girlfriend's name on their arm. Spiritual concerns, too, were mostly discussed in terms of opportunities to see families. Pchum Ben, a popular Buddhist holiday when the spirits are supposed to come close to the living, was a time when people talked about visiting their home and seeing their relatives or the landscapes they missed.

Most of the day, the platoon worked under the concrete Buddha statue without using religious terms unless they were talking about love for the rats. Neither Buddhism nor spirits seemed particularly important. Despite ghosts of war being a large part of postwar ecologies in Cambodia and the rest of Southeast Asia—embedding landscapes with *barami* (power) such that certain places are considered good for travel, planting, or love—the ghosts of war seemed to have no effect on the postwar ecology where I did my fieldwork (Guillou 2017).

That is, until a ghost killed Håvard, the rat.

.   .   .   .   .

Håvard was one of the rats in Moch's unit. Each pair of deminers had been assigned two landmine-detection rats to learn the landmine-detection-rat technique. Moch was younger than Narin, who had over seven years of experience in demining and before that, combat experience. The supervisors had also paired the deminers so that one was a government employee and one was an NGO employee in order to mitigate conflicts and suspicions among these competing teams. Moch had always been kind to everyone, and she had never had any problems with Narin, who had taken on a paternal dynamic with her. With a soft face and generous smile, Narin cooked with Moch even though people made fun of him for cooking because he was a man. They were a quiet pair who were always respectful toward each other, despite the rivalry of the NGO and the state organization.

Håvard had been biting his two human handlers a lot. The rats had been assigned deliberately: each pair of humans had one rat who would be "difficult" and one rat who would be "easy." Although all the rats had been selected from Tanzania as relatively easy rats for their implementation in a new country, there were still relative variations among them.

Difficult meant different things: a rat might be difficult if he did not move quickly in the pit to sniff for TNT and groomed in the middle of work instead; if she chafed at her harness and tried tugging it off; if he ran away from the deminers and did not go straight into his transport cage when he had completed his detection in the pit; or if she bit fingers and hands when being held. *Easy* meant that the rat followed her humans when the humans tapped on their legs, that he wore his harness easily, and that he worked well in the minefield pits. Moch preferred her easy rat, Simon, whom she considered "the most proper rat and very clever," using words associated with the politesse and diligence expected of young children at school.

But even though Håvard gave Moch and Narin difficulties, they both caressed him gently when smearing his ears, tail, and feet with sunblock. Even if he bit their fingers, the two would only growl their annoyance, as if at a child, and then continue to carefully apply the lotion. One week was particularly bad and both Moch and Narin had Band-Aids on the tips of three fingers. Moch would never talk critically about Håvard, but she favored her other rat, Simon, who would snuggle his nose into her shoulder when she lifted him and held him loosely until he crawled up her bicep. "Simon is a good and proper rat," she said as she stroked his back.

The next day, Liz, the British supervisor who had been hired to ease the implementation process in Cambodia, announced her resignation to the platoon. She and her partner wanted to return to Laos to live. She enjoyed landmine-detection work and considered it essential, but her heart was in the other half of mine action—that is, amputee rehabilitation. After the landmine rats were trained, she would be starting another job with Handicap International in Laos.

She held a party on the minefield to say good-bye. It was midmorning and the training had been cut early. The rats had been brought back to their kennels and now huddled in their terra-cotta pots to sleep the rest of the day. Outside the kennels, behind the Buddha statue, we all sat on the tiled steps. I sat next to Moch and Sovannah, who were giggling with each other. Then, Liz got up. "All right, all right," she said, interrupting the revelry. Chann stood next to her, ready to translate her English. Liz had a commanding presence and everyone stopped to listen.

"I know that I'm leaving," she said, "and I just want to say how proud I am of all of you. And how these rats that we have here are real heroes."

Chann translated, and people nodded after his Khmer.

"And I'm going to come back and visit. And these hero rats, they will die."

This was a weird moment. She paused after the word *die* to wait for Chann to translate, but even Chann, who often played practical jokes and teased people, frowned when she said that. I'm still not sure why she mentioned it.

"When they die," she said, "I want the rats to be memorialized properly. I want them to have a headstone, and the headstone will say how many bombs they've found."

Chann's translation had people suddenly stiffen. Then Liz continued and spoke again of how proud she was of the deminers learning a new skill. The party continued, too, and we finished the day with a board game played on the tiles of the CMAC room. No one mentioned to me anything about Liz's speech.

The next day, I received a call from Chann before dawn. He couldn't pick me up, he explained, without telling me why. I was happy to sleep in, instead of waking before dawn to ride my bike to the Siem Reap offices and get a ride to the minefields. Liz and I were supposed to have one last interview for my research. At 8:00 a.m., after eating breakfast, I received a text message from Liz: "Meet me at the Siem Reap office."

I was already dressed. The morning sun in Cambodia was bright and woke me early. I grabbed a tuktuk to the Siem Reap office.

"What's the matter?"

I met her at the government demining compound in the city, walking down a muddy river path to the concrete mansion, where she was smoking outside and frowning. She shook her head when she saw me. "One of the rats is at the vet—we have to go." She pointed to the car, and then I followed to the passenger side. She didn't look at me.

"Which rat?" I asked.

"Håvard."

"Oh! He was biting a lot last week—maybe he's sick?"

She didn't say anything as she navigated the jeep along unpaved narrow roads to the river where there was a high-end shopping area. As we drove along the river, she said, almost to herself, "I bet he's already dead. I bet they're just not telling me."

Death itself is an unspeakable thing for my Cambodian interlocutors. The power to speak of a person's death in some ways seems to be understood as possibly causing it (Rechtman 2021). Erik Davis (2015) calls this "deathpower," in a play on Foucauldian biopower, and Mbembe's term "necropolitics." Davis's ethnography shows how elite players who compete for power in Cambodia struggle over "caring for the dead"—that is, spirits—because it relates to "a modality of control over the dead" (2015, 5–6). As such, spirits with deathpower are not directly spoken about, even though Yeay Mao and other spirits are everywhere in Cambodia. They are part of the *neak ta* or *neak yeay*—grandparents or ancestors who are fiercely nationalistic about Cambodia, connected to its land and its history, and who are liable to take revenge on worshipers who do not give them respect in the form of gifts and sacrifices (Jacobsen 2008). Instead, things like Facebook photos offered my interlocutors ways to tell me stories about ghosts and spirits, which in turn allowed them to communicate fears about them. Photographs would often allow for the reality of both the spirit and the violence that would normally be unexpressed. Like the "visual evidence" of the skulls in the glass tombs of the killing fields, "images function as tools to facilitate the discovery of truths not stated or known in the creation of the images themselves" (Lim 2016, 4). This mode of communication encompassed the reality of some of my interlocutors' shadow stories, such as the transgressions Chann and Hien sometimes hinted at. But there were no photos of Håvard shared with Liz and me.

We drove to the city's only pet store to see him in person—a small boutique that catered to foreigners. It had windows decorated with stickers of skinny cartoon women holding leashed poodles. The door was open. Past aisles of plastic cat toys and brightly colored bags of brand pet food, the vet, Tokla, stood at the back of the store looking down into one of the rat transit cages on a filing cabinet. It was transparent, lined with paper towels and the size of a shoe box. Heng, Liz's assistant, and Hien were there. Chann was still at the minefield.

As we walked in, we could see Håvard on his back in the cage, mouth partly open and legs curled tight to his white belly. Tokla, the rat veterinarian, had tears in his eyes. As soon as she saw Håvard, Liz left the boutique to call Tanzania. Tokla told me in Khmer that Håvard had died in the car. He mimicked Håvard's last noise, a gulping gurgling cry. We both looked down at the rat's body, early morning light flooding in from the glass doors where Liz was pacing and talking rapidly on her cell.

"Ask her if we can do an autopsy here," Liz said, interrupting her international phone call to the headquarters in Africa. The woman in the store looked sympathetic but explained that there was no vet there who could do an autopsy. The vet at the Siem Reap hospital who did surgeries was visiting New Zealand.

We all drove back to the minefield with Håvard's corpse.

At the minefield, repeating the scene of the pep talk and farewell address the day before, Liz gave another speech with Chann translating. She told the platoon that they had to make sure the rats were resting enough—"If they're tired, then just let them rest." She told them that they were going to drive Håvard six hours south to Phnom Penh to conduct an autopsy to see why he had died.

Håvard had been Moch's rat. Moch was weeping silently. She stared at the ground. After Liz's speech, Chann began to talk, not translating as usual, just beginning to explain something. Liz just stared at him with her eyes wide. She did not understand Khmer, but him speaking was strange as well as a bit of an overstep.

"I think also, because of what Liz said yesterday about a cemetery, that we haven't paid enough respect to the spirits, and the spirits heard her and now we have to pay them some respect." Liz glanced sharply at him at the sound of her name.

Chann and another deminer started talking about this earnestly, agreeing that they hadn't paid enough respect to the spirits. He used the general word for spirits, not specifying if it were the grandparent spirits that protected certain lands or the bad death spirits that haunted places where they had died. The "spirits" here seemed to be diffuse, human in form, but unnamed in terms of how much power they had. And so Chann sent one of the younger deminers to the village on his moto to get incense and a ribbon and some fruit for the spirits.

"What was that, mate?" Liz asked, hearing her name and seeing the man leave. Chann told her. Liz didn't say anything. She shook her head again tightly, barely perceptible, and then began to start her multiple phone calls across time zones and continents to figure out what the NGO leaders wanted to do.

Even I had trouble doubting the causal effect Liz's words had had. The spirits, Chann told me, "always listened." I imagined, then, that the awkwardness I felt the day before during Liz's speech had to do with Liz mentioning death when everyone else knew that spirits were listening. The timing was perfect—Håvard had died because Liz had said that the rats would die and this angered the spirits. It was unclear to me why the spirits would be upset, and when I asked, people would just say, "We haven't respected them enough." It seemed as though Liz's words had reminded the spirits that they had not been properly respected and that they could cause a rat's death.

The younger deminers on motos brought back fruit and incense for everyone. While Liz and a few others drove to Phnom Penh with Håvard's body to have a vet there perform an autopsy, I stayed with the platoon to kneel at the Buddha statue. I lit incense with everyone else at the altar in front of the minefield where Buddha smiled over the sandy dirt.

Håvard's death revealed the presence of spirts on the minefield, and it also opened a door for the deminers to begin talking about spirits. Prior to this, stories about spirits had been rather scarce. When I had first entered the platoon, I had told Liz about my interest in spirits in the countryside and ghost stories. I told her the stories people had told me about the various killing fields where spirits needed to be put to rest by being displayed in glass-walled pagodas. She had told me, "Yeah, in Laos and Thailand, the spirits seem like a big deal, but here Cambodians don't seem to mention them at all." She had lived for decades in Southeast Asia and had seen no evidence of spirits in daily talk in Cambodia.

But when deminers had been asked about spirits or the shamanic magic they used to protect themselves in the minefield, every deminer seemed to have different stories about someone they knew who had yantra tattoos that allowed spirits of protection to enter their body, or pouches that they wore around their waists, or stories of ancestor spirits that visited them in their sleep. On the minefield with the platoon and the rats, these stories increased after Håvard died. Offering the spirits on the mine-field a bowl of fruit and incense once a week through the statue of Buddha seemed to be a solution to the threat the spirits held over the rats.

Although Liz had fretted about a disease that the other rats could catch, the autopsy came up inconclusive. The inconclusive autopsy provided more evidence that a spirit had killed Håvard. Its inexplicability became evidence of the spirits. The deminers hoped that the offerings would ensure that no other rats would die, but I was confused about the talk of spirits when the bowl of fruit was being offered to Buddha. I asked Chann about this after we offered the spirts a bowl of fruit and lit some incense.

"How do the spirits get the fruit if we're giving it to Buddha?"

"It's the same. If you give gifts to Buddha, you give gifts to the spirits, too."

"But the statue is just Buddha?"

"All the spirits from all the world are in the statue. It's not just Buddha."

We looked at the serene smiling face. Usually, Tokla slept in a hammock beneath the Buddha statue because he preferred to sleep outside, but his hammock had been taken down because Chann had insisted that no one could sleep on the steps of Buddha.

"Even my grandmother?" I asked.

"Yes, yes, your grandmother, too."

I felt satisfied with this answer and, although surprising in practice, it seemed in keeping with what other scholars of Cambodia have noted in terms of the distributed personhood of spirits as diffuse and mutable (Choulean 1988; Davis 2015; Arensen 2017).

Chann continued with a comment that raised more questions about his cosmology, which presumably Tokla and Hien shared, because they listened in and nodded after he said, "Buddha is like the manager of all the spirits."

When Chann used the term *manager* for Buddha, he evoked the employee tree at the front of every work office, the organogram. These employee diagrams looked like a family lineage, with the head of the organization at the top, the manager, and the employees in descending lines. Chann went on to draw this analogy more explicitly, explaining the pyramid structure of hierarchy that was also reflected in the spirit world. It was a mark of prestige to be at the top of the tree, but calling Buddha a manager spoke to a kind of lack of differentiation between the spiritual world and the human world.

This was probably striking to me since my own categorical distinctions enforced a strict separation between the two worlds and, in fact, is often upheld as a moral imperative. For instance, Christian doctrine often mentions a division between worldly and spiritual powers and considers the pursuit of worldly powers to be a sin. In other places in Southeast Asia, however, conflating of these two categories was a common way to understand the spiritual world (Tambiah 1984; Forest 1993; Kwon 2008; Zani 2019).

On the minefield, I found that these worldly/spiritual conflations reflected the materialities of the bombs and had ramifications for understanding how to deal with the dangers of violence, relating directly to the fears the deminers had about their coworkers and the state. The spirits, like the state, could hurt you if you did not respect them, but they also could protect you. Amulets, one form of this spiritual protection, make the connections between physical materialities and spiritual realities explicit. Alain Forest (1993) describes an amulet made of a burned fetus taken from a pregnant woman. The amulet holds the spirits of the woman and the fetus, and the spirit of the woman will do anything to protect the baby. Several deminers I knew repeated stories about such amulets. For the deminers, spirits and Buddha were not separate but were intimately tied to worldly powers. They were material beings that interacted with humans and nonhumans through inexplicable sensorial encounters.

.   .   .   .   .

When the deminers spoke of love for the rats, they often evoked Buddhist philosophy through words like *metta* (which I explore more fully in the following chapter). By calling them "little sisters" and "very best friends," they revealed an affection for them that also referred to the *samsara*, the cycle of reincarnation where people become reborn as other living beings and vice versa. More abstract Buddhist philosophies played out on the ground in everyday common practices. For example, philosophies of impermanence in Buddhism become interpreted by villagers as an attitude of accepting hardships as they come (Cassaniti 2015). Rats were not just technologies but also nonhuman persons. I have already described the beings of the rats in terms of their sensorial encounters and their relationships with humans, but the bombs presented different kinds of actors. While the rats and humans were clearly related via materialities and shared sensorial encounters, the landmine's sensorial encounters were more a part of the ghostly realm because they were incomprehensible, a sensory encounter that undermined stable categories of being human.

Key to divisions of destruction and production are divisions between life and death that feminist scholars have undermined through their empirical work (Lyons 2020; Govindrajan 2018; TallBear 2017; Mbembe 2019). Scholars of Southeast Asia have also moved toward considering hauntings and deaths as inseparable from categories of life on the ground (Klima 2009). Heonik Kwon (2008) describes how the ghosts of soldiers and victims alike haunt the former battlefields of Vietnam. These ghosts take on political imaginaries, serving as potential alternative realities for people who seek to imagine ideas of peace rather than war. For instance, he describes how a well-known spirit, when speaking through his medium, did not understand or remember the war properly, and this lack of memory from the spirits presents a spirit realm where former enemies are no longer enemies.

Landmines and other military waste remnants in Southeast Asia collide with these hauntings, as Leah Zani (2019) has pointed out in her ethnography of Laos. The ghosts of war in Laos haunt alongside explosive remnants of war (ERWs), which presents a land contaminated with ERWs as a generative landscape both conceptually and literally. The ERWs in Laos and Vietnam evoke war but also allow civilians to imagine parallel worlds. They also often keep land safe from development so that land generates lush landscapes. In the case of Vietnam, Kwon (2008) describes the spectral worlds of the ghostly soldiers and victims of the civil war as potential alternatives to the current Vietnamese state, alternatives that simultaneously evoke unity after the war and subversion of the powers that be. This follows with Eleana Kim's descriptions of landmines in the DMZ, which I described earlier, and Zani's depiction of how a spiritual land of peaceful "fruit eaters" offers a parallel of the Laotians who salvage ERWs that they name after fruit that they can pick. These parallels show that the materialities of war effect conceptual and ghostly realms all at once. As a sensory ethnographer, I examine these spiritual encounters as a kind of inexplicable sensorial encounter. If my interlocutors told of their encounters with ghosts, they represented them with stories characterized by sensorial eeriness. When I asked about my interlocutors' beliefs in spirits or ghosts, the answers often were contradictory stories, such as a woman who told me, "Ghosts don't exist, except once a beautiful Chinese woman appeared to me and I knew she was my ancestor visiting me."

I frame sensorial encounters on the minefield as material encounters. In doing so, I am putting myself in conversation with sensory ethnography schools that ground imagistic knowledge in sensoria and phenomenology, but in the context of my field site, these sensoria varied according to being, capacity, and world. While perhaps we can imagine what it is like to be a rat by testing our sense of smell, being human limits that sense in that we cannot smell as a rat does—rats encounter materials in ways that we cannot but can share only through their indications. One scientist described a rat's capacity for scent detection to me by suggesting it would be as if humans could smell colors. It was typical that a human being prioritized sight as an analogy, but the point was that a rat's sensorial encounters are not beyond human comprehension, only beyond human capacity. Ghosts, on the other hand, present *extrasensorial* encounters—a portmanteau of *extraordinary* and *sensorial*, not *extra* as in excess. The stories about them are no less material than others and, indeed, ground themselves in death materials, as postwar ecologies. Although natural versus supernatural and material versus immaterial are common dichotomies to characterize these ghostly realms, *extrasensorial* grounds such realms in the world of beings. Mayanthi Fernando (2017, 2022) depicts the difficulty of grasping this, suggesting that agency must be rethought through porosity and phenomenology.

I want to think of these Cambodian spirits as real—that is, as materialities and beings. Although anthropologists often present spirits as social constructions that serve a purpose for political resistance, for people to cope with ever-present signs of violence or for citizens to rebuild national identity and community, I want to say both that the ghosts are interpreted as signs and that they interpret things as well (Kwon 2008; Singh 2012; Uk 2016). The spirits' actions change depending on the meals they are given, and they possess and act through various Cambodian bodies and artifacts. Thus, while spirits "emerge with human practice," they are "not reducible to or circumscribed by the human contexts in which such practices unfold" (Kohn 2013, 280). When a woman is possessed by a *neak yeay*, as in the case of a Cambodian garment factory worker, it is her grandmother, an ancestor spirit speaking through her, not herself.[2] It is a different kind of self, one that is shared in a "continuity of selves" (Wallace 2014; Kohn 2013, 265).

On the minefield, the spirits were described as a diffuse power within a hierarchy. Buddha, as Chann told me, was the "manager" of all the spirits, but with a few exceptions most other spirits lacked this level of individuality. Spirits are tied to war and are also beings crucial to understanding relations after war, but they also, as has been pointed out by other scholars of Southeast Asia, represent modes of governance and how surveillance can be both harmful and beneficial to the ones being watched.[3]

As hierarchical as spirits are, they portray ways of understanding personhood as unfixed, "dividual" rather than individualized. Although they often have names and even personal histories, these spirits mutate due to their changing environments, needs of their human interlocutors, or even their own ambitions. This is common throughout Southeast Asia and presents another example of how personhood can be flexible and mutable.[4] A famous rotary statue in Siem Reap was originally sculpted as Buddha, but, for at least as long as I have been visiting Cambodia, it has been worshipped as a grandmother spirit, Yeay Mao. She sits, like Buddha does, covered in golden sashes, but her face is often decorated with lipstick. A statue of a former Vietnamese soldier near Pol Pot's grave has also been transformed into Yeay Mao (or a site for worshiping Yeay Mao) through the same kind of physical modification. Humans offer both these statues bananas as a nod to Yeay Mao's tendency to consume phalluses. Other spirits have been asked to transform for human needs. For instance, in Phnom Kulen, a grandmother spirit once took care of a field, but she mutated after requests that she keep villagers safe from accidents. She then was renamed Grandfather Safety (Arensen 2020; Sakhoeun 2020). Rather than thinking of this as animist, as a spirit who inhabits or possesses a statue, the worshipers I spoke to see these mutabilities as physical and spiritual transformations that occur because of the diffuse personhood of spirits in general.

Figure 7. A spirit house for the Demining Headquarters in Siem Reap at dawn. Photo by author, 2015.

While their bodies and persons are subject to change, the spirits are not completely distinct from the human world. They metabolize food and they have material desires. Most homes have a miniature, baroquely decorated "spirit house" in front of them. To keep the humans at home happy and wealthy, the family fills little glass votives with tea and red Fanta, lights incense, and offers fruit, candies, and other treats that the home spirits eat (figure 7). The spirits consume food from the material realm with similar desires and tastes as humans. At the same time, materialities take on inexplicable sensorial encounters for the spirits. On several occasions, when I was admiring a spirit house, I was told a "joke" about its offerings:

"You know why the spirits drink red Fanta?"

"Why?"

"Because it looks like blood!"

The material world acts as a funhouse mirror for spirits—they need to be fed but they can be tricked into thinking a red soft drink is living sacrifice, meaning that taste does not work the same way for them as it does for humans, that the spirits have different kinds of sensorial relationships to materialities than humans do. These sensorial encounters alter the human world, too, sometimes with violence, hinted at in the red Fanta joke.

.   .   .   .   .

Talking about the spirits was a bit taboo. Like some Cambodians managed suspicions and fears from memories of war with shadow stories, they sometimes made the spirits subjects of black humor (albeit as respectfully as possible). The spirits also seemed to give shape to fears of destruction and state violence. But they were also more than simply metaphors for the state and the fear of minefields. They were material beings that could kill a rat or protect you from a landmine. They needed to be fed and respected. Yeay Mao was particularly popular in Cambodia. As a being, Yeay Mao took on the form of an old woman but also existed as a young woman, depending on the statue. She protected travelers and took revenge on husbands who were unfaithful to their wives. I heard about her quite frequently in the markets and from women, but in the platoon most of the deminers were men. It was only during the telling of a sensorial encounter that I realized Yeay Mao had a darker side.

My Cambodian friends constantly shared graphic images of violence or sex on social media. Despite being fairly close-mouthed about tragedy, death, and violence, photographs on Facebook allowed them to share horrifying news and imagery portrayals of horror—the images were connected to no individual author but rather to a collective, which allowed for graphic depictions that sidestepped the taboo of an individual authoring the image. Sexual and violent images emerged in a Cambodian user's newsfeed without any fault from the user; perhaps this is why nude photos and sometimes links to porn videos and animations were often shared out in the open on Facebook, and sometimes even through official government profiles. While modesty norms generally enforce shame over nudity (Uk 2016; Ledgerwood 2008; Jacobsen 2008), social media allowed some Cambodians to express what was normally considered taboo.

My newsfeed was thus rife with violent images most of the time. These images would be considered too "private" or "graphic" for public circulation for most North Americans or Europeans I knew, who often complained about the images shared by their Cambodian networks. The slew of photographs disturbed me, too—my partner at the time, a Canadian, often spoke about being in a bad mood because he had seen too much of "Cambodian Facebook."

I imagined there was little that Cambodians wouldn't share with me until, in the fall of 2015, my interlocutors were reluctant to show me the photos they were looking at during lunch time. We were at the restaurant Morning Star, and as usual, Hien was on his iPhone checking Facebook. While we were eating, he stopped and widened his eyes. "Look at this," he said to Chann and passed it across the table. Hien tutted and shook his head, but then said, "Yes, it's very cruel."

"What is it?" I asked.

But Hien passed it to the other men at the table, two of them the new African consultants, who rolled their eyes at the photos. "Come on, guy, don't share that." We spoke in English because these men did not speak Khmer. One of them said, "Hey, not while we are eating."

Hien took his phone back and he shook his head, too. He said, "Oh, it is terrible. A man was killed."

At this point I had already guessed it was a picture of a corpse—these were often shared on Facebook feeds in Cambodia and on the daily television news. I had already seen a video of a twitching body under the wheels of a truck on the television's morning news while I ate my breakfast. On Facebook, I had been shocked by the photos of murdered women lying naked on the side of Cambodian roads. These photos made me shudder, but I still wanted to see what was on Hien's phone, so I said, "Oh, is it a car accident?"

This was a good guess because fatal traffic accidents in Cambodia numbered in the thousands per year. I was surprised then when Hien shook his head no. And when he showed me the photo, I realized that the image was particularly gruesome and that was perhaps why they kept it from me—a man's face in profile stuffed with a dismembered penis that flopped from his mouth was pictured on the screen. My eyes shut at the sight of it and I whined a little. I tried to forget the tint of purple on the corpse's flesh.

"Yes, it is awful, just awful," said Hien, and he took the phone away.

"This was in Kampong Chhnang," said Chann.

"Yes," and Hien waved his hand around. "This man's wife was angry because he had a secret girlfriend and so she cut his thing off."

"And he died?"

"Yes, he died. Of course he died."

And then Hien showed us another photo of a man murdered in the same way. "This is in Thailand."

"It is like Yeay Mao," I said, referencing a spirit who murdered men by eating their penises, and Chann nodded.

"Yes, Yeay Mao does this," he said. "She protects women by killing their husbands who are cruel. When they have secret women."

Yeay Mao was a goddess, a woman who died searching for her husband who was lost in the war. In stories about her, this war was never historically defined, but it was said to be against the Thai and before the colonial period (Jacobsen 2008). Before 1944, she was unappeased by offerings, causing trouble to passersby and road accidents. After 1944, she was considered a potential force for protection, especially for women. When Chann and Hien mentioned "secret women," they meant an affair that had been hidden from a wife. If a husband was unfaithful to his wife, Yeay Mao, the grandmother spirit, would cut off his penis and eat it.

People worship Yeay Mao and prevent her from consuming men's penises with phallic-shaped offerings like bananas. There are two versions of her: one that inhabits a statue of a young woman on the Kampot coastline—this one is called Chomteav Mao (Lady Mao)—and the other, Yeay Mao (Grandmother Mao) lives in a huge statue on the main road from the neighboring province. The statues manifest the spirit in different forms at the borders of the province, both coastal and inland, to grant the province protection (Kitagawa 2005). I have visited this spirit in the form of her twenty-nine-meter statue, erected in 2012, and in the white statue that sits like a mermaid at the coastline. I have heard many different and sometimes contradictory stories about Yeay Mao. For instance, she controlled armies against Thai invasions after her husband, a warrior, died, but she also was said to be a woman who was killed in the ocean because she was searching for her husband and since then has hated men. She was also said to be a warrior woman whose husband left her for a princess, and she went mad with grief and promised to kill two million of the country's strongest men.

Yeay Mao is a very powerful spirit who is associated with Kampot and Kep provinces in the southwest of the country, although she is popular everywhere. Like other *neak ta,* Yeay Mao is a spirit who has had a "bad death." Anne Yvonne Guillou (2012) connects these spirits, usually former warriors, to the ghosts of Khmer Rouge soldiers who sometimes exact revenge on "careless human beings" (221). She explains that the *neak ta* allow Cambodians, especially rural Cambodians, "to express social suffering, but also to give local meaning to the genocide" (226). Guillou suggests that even Pol Pot has become a kind of *neak ta* (she does not name him as such, referring to him only as part of the "Pol Pot dead" who communicate and grant power to those who worship them) and that these ghosts offer ways to remember and connect contemporary violence with that of the past (224). Yeay Mao's name refers to her as a grandmother, *yeay,* but she is also a war spirit. The settings are extremely important for these spirits, and they become ways to remind the living about the events of the past that imbue the land with power. Yeay Mao, similarly, allows for a connection between photographs of murder victims and her own participation in wars and gendered violence.

The photos Hien circulated showed two men with their phalluses prepared as food, and when I connected the deaths to Yeay Mao, both of them elaborated on my speculation. Like the victims of Yeay Mao, according to Hien and Chann, the men were murdered by vengeful women. Hien and Chann both explained that the victims had cheated on their wives. They laughed at this and said that this was why you could not have secret girlfriends in Cambodia—"because of Yeay Mao."

Yeay Mao has power over the two dead men in the photographs and, as such, has power over Hien and Chann who might die if they are "cruel" to their wives. Because of her power and because of her surveillance, even "secret women" are not really secret.

Although Chann and Hien never directly told me that they cheated on their wives, they joked about it. They even offered me a boy when they hired young girls to join us for beer and supper. And here again, they joked about how men who cheated on their wives should be worried about Yeay Mao. The photographs of these murder victims allowed them to become potential victims via the real actions behind their implied affairs.

As we continued our meals, they did not directly mention Yeay Mao again, but the photos and the murder story inspired Chann to tell more stories about ghosts who had deathpower. Chann asked us if we wanted to know another ghost story. I found myself mostly quiet, shocked by the photos of two dead men with penises in their mouths. But he began to tell us the story of a woman who fell in love with her daughter's betrothed. Her beloved was less than sixteen years old. When she told her husband, he was so appalled that he told her to go away, but she wouldn't. And then the older woman and the young betrothed conspired to murder the husband. That man was killed and the two abandoned the teenaged girl.

"She was all alone in the village, then," Chann continued, "and because she was so young, no one could make the proper funeral arrangements for her father. So his ghost haunts that house—she left to live in a neighbor's house and now only he lives there. You can still see it today in Kampong Chhnang."

In fact, Chann was so certain of this that he gave me detailed directions to the house, even when I insisted I was afraid of ghosts so I'd rather not visit.

"But so why doesn't the girl take back the house?"

"She was all alone, so she had to stay with other relatives. She was too young."

"Very cruel," he commented, a phrase he often repeated when talking about animals or humans and violence associated with them. In this case, the vengeance of women resulted in a tragedy for another young woman. I wondered about the quick succession of stories. While the violent vengeance in the first stories rendered the women heroes and the husbands the villains, in the next story a woman's violence rendered her the villain and the murdered husband and daughter the victims to be pitied. The stories together seemed to tell how infidelity resulted in tragedy no matter who the unfaithful partner was—although, in both cases, a man was murdered. By the end of the story, a young woman is left alone.

Hien continued to look at the stream of corpses that began showing up in his "Related Stories" section. The corpses of the men just lay on the ground, and some of the photos were more extreme close-ups of the men's mouths. Out of the side of my eye, I saw things resembling slugs flopping over their lips. Hien shook his head. "Very impressive," he said, "that a woman could do this."

The death as an "impressive" act indicated a sense of the inexplicable when it came to their murders. "Yes," Chann agreed, "very impressive."

How could a human woman do such a thing—overpower her husband and cut off his penis to stuff it in his face? Calling it impressive seemed to be about physical capacity rather than moral capacity, although this was part of the conversation, too, with Chann repeating, almost wistfully, "Very cruel, very cruel."

This is when Hien and Chann directly linked this murder to an extra-sensorial encounter—a woman who normally should not be able to over-power her husband did. Both the ghost stories and photographs left open the possibility that spirits had punished the man's infidelity themselves. Hien's compliment to the women behind the acts as "very impressive" also expressed doubt that a woman could do that kind of damage to a man. This could imply that she didn't work alone and that men or even that spirits, probably Yeay Mao, had helped in the crime. Not only did the pho-tographs depict unspeakable things, like death and murder by spirits, but they also indirectly represented anxieties about spirits, gender, politics, and a past full of atrocities.

Yeay Mao and other *neak* spirits signify wars and violence. They are intimately connected to the hierarchy of power that includes politicians and monks. Yeay Mao herself is a spirit who is both vengeful and protec-tive. She is a veteran of war. She represents a person who could be both ally and enemy. She renders relationships and fixed categories of good and bad unstable. At a place in Anlong Veng, Yeay Mao enters the statues that memorialize Vietnamese soldiers, leading worshipers to dress them in scarves and put lipstick on them. The soldiers, who are understood in many contexts to be enemies, suddenly become a protecting saint. I learned that ways of talking about Yeay Mao and ghosts relate to ways of talking about violence and that sometimes spirits offer ways to portray violent war stories in unexpected ways.

·　　·　　·　　·　　·

Even though the spirits were a potential risk in the present, Hien told me stories about how spirits had helped him, which was why I shouldn't be afraid of them. After Håvard had died and we had held a ceremony where we gave the statue of Buddha fruit and incense, Hien and I sat in the shade waiting for the deminers to finish their training. Hien was not often in the minefield because his duties kept him in the city, but Liz, Chann, and another supervisor had gone to Phnom Penh for Håvard's autopsy. So Hien hung out with me to practice his English over the next few days.

At ten in the morning, the sun was glaring so much that we needed to rest. He and I sat with the rats, who slept under the shade of the trees in their plastic transport boxes. They slept with twisted bodies, mouths partly open and eyes closed. They reminded me of the dead Håvard. Merry, as usual, whimpered and whistled in his sleep. Hien speculated that on that day Merry was remembering his dead friend.

"Don't worry, Darcie," Hien said. "It will not happen again. We will give the spirits fruit once a week. I will buy it for them. If we respect them, it will not happen again."

I shook my head, telling him that I was a little afraid of spirits.

Because Hien saw my worry about the spirits, he told me not to worry. He said that he was born with a placenta around his neck and his face and because of that he was safer than other people from the spirits. This safety, in the form of watchful spirits, also attracted increased risk. Lightning, for example, was attracted to him. When he was in the rice field, he said, he had been struck by lightning three times in his life. "Just in 2013, I was alone in a rice field and I got struck by lightning! But nothing happened. I just fell down and then I was fine."

Hien compared this lightning to the ordinary visits from the spirits. "They tingle all along my arm when I talk to them," he said as he pattered his fingertips down his arm.

Hien tilted his head and leaned back on his big arms. "When I was younger," he began, "my grandmother and I were escaping the Khmer Rouge—people were leaving, a whole train of cars. On the car's path there was an antitank mine that blew up! Whewww! All of a sudden, I was flying in the air. My grandmother did, too." As he described his younger body flying up from the force of the explosive, his hands drew an arc in the air from the starting point of the ground to three feet above it. He blew through pursed lips to mimic the sound of air rushing all around him. Then, rather abruptly, he said, "I landed. Where was I? I was looking for my grandmother. I couldn't find her. But I was fine because the spirits watched out for me. I found her, and they had saved her, too. Oh, yes, it's normal, my grandmother told me. I was born different than other people."

Hien did not finish the arc—did not allow for his hand to fall to the ground and complete the child's fall from the explosion. Instead, he interrupted himself, leaving, in his words, "the bad thing" he remembered unspoken. The story instead developed into a story about his "friends," the spirits—"I was fine because the spirits watched out for me." Later he said, "They are my friends." The explosion and the escape from the Khmer Rouge began to retreat from view, but they were still there, waiting in the background.

In addition to offering safety, though, Hien's shadow story made clear that the spirits also bring risk to those who pay them homage, as they brought risk to Hien in the form of lightning. This goes for people who are not "born different," too. The spirits are very dangerous. For me, this was interesting because the dangers the spirits posed were similar to the dangers the state posed. They both watched the minefield. They both could offer some form of protection—like the state car the deminers used, whose CMAC logo ensured that no police officer ever stopped it for tolls—and like Hien being safe from a landmine because of spirits. But by that same token, they both were potentially violent actors and their actions were uncertain and unpredictable.

Hien and other deminers made this comparison between spirits and state explicit. Hien explained: "I am not afraid of anybody. I give them all respect, but I do not fear anybody. Not CMAC officers, not spirits. They will support me and they are my friends. You shouldn't fear them." CMAC officers and spirits here are placed in the same category of beings who need respect but not fear. This was like understanding Buddha as a "manager" of the spirits. The spirit world became an indirect lesson for how to engage with state powers and surveillance.

When the deminers buried Håvard, they followed Liz's instructions. Although the vengeful spirits who watched over the minefield in wait to murder a rat were diffuse—to give Buddha fruit was to give them all fruit—Håvard received an individualized funeral and tomb. Giving fruit to Buddha in care of the rest of the spirits had a vagueness that gave the actions to appease the spirits a kind of just-in-case, precautionary feel. To honor them diffusely is to honor every one of them, so that none would feel more or less honored. This contrasted with the very specific honor granted to Håvard, the first rat of the Cambodian landmine-detection-rat platoons to die.

Håvard had a shrine of his own. A deminer who lived near the minefield had volunteered a tiny plot for his burial. They had cleared the plot and poured cement into a hole, forming a tomb. The body had been placed into the cement walls and then it had been buried. I remember staring at the rectangular hole where the curled up rodent body had been placed. His eyes were closed and I remember thinking about how his paws looked like praying hands. Atop the dirt, a tombstone had been placed, and deminers and I posed next to it for photographs. We smiled in the photographs. He had not yet detected any active landmines, since this was before the rats were certified to work in active minefields, but his tombstone still had his name. Liz had gotten her wish.

My friends had invited me to pose next to the tombstone in honor of Håvard but also to honor the generosity of the young man who had offered this little plot for the dead rat.

I end this chapter with this rat burial because it contrasted with the cosmology of diffuse personhood that the Cambodians presented to me. I parallel this with the ways in which the skulls housed in the glass pagoda represent a compromise of two opposing memorialization systems. The memorializations of these killing fields, skulls that stare out from behind the glass, represent a compromise between two competing and contradictory ideas about the genocide and its violence: an archive versus an impermanence. The unclaimed bodies from the Khmer Rouge genocide with their restless ghosts can be laid to rest in the Buddhist cremation ritual, but then the bodies disappear, which from one perspective can be read as burning evidence. On the other hand, burning the bodies aligns with traditional Buddhist ideas of impermanence, where it becomes sacred to forget "to remember again" (Thompson 2013). Such ideas of impermanence fit better into a cosmology of diffuse and mutable ghosts than does an individualized memorial of a singular tomb with a rat's individual accomplishments.

But this was not only a Buddhist value of impermanence at play. I connect this different aspect of memorialization to a different construction of relations embedded as part of a postwar ecology. Individuality is emphasized in the way the rat is remembered, whereas relationality is stressed in the way the spirits are honored. In the context of the minefield and given the history of violence in the pasts of most of the ghosts, this relationality connects to ideas of rebuilding relationships after a civil war.

When thinking through concepts of justice and reconciliation, we can see that global models rely on the clear separation of enemies and individual perpetrators and victims. Such categories often do not work on the ground, as anthropologists and other scholars have pointed out. In her critique of the Khmer Rouge tribunals, the legal scholar Virginia Hancock (2008) suggests that tribunal forms of justice could fail Cambodia. For them to succeed, she recommends that the tribunals account for the fact that Buddhism emphasizes a "community-oriented theory of crimes against humanity," in that the judges should not understand harm as involving only individual culprits and victims (88). This individuality, she suggests, does not consider the modes of resilience enacted by Theravada Buddhists.

Who can be held accountable for violence in Cambodia's postwar ecologies if everyone is at once perpetrator and victim? If personhood is truly diffuse?

Extrasensorial encounters offer a bit of an answer to this question of shared accountabilities. The dead victims of the genocide seek retribution through the peaceful interment of their bodies. Yeay Mao, a former warrior, needs to be fed, and the spirits keep people safe as long as they are "respected." These spirits exist in a hierarchy of power much like the hierarchy of power of the state. The respect for the spirits can be seen as a model for how to navigate worldly powers: respect for CMAC officers meant that Hien had nothing to fear from them. But it also embedded Hien in shared obligations—both he and the CMAC officers had to respect the spirits. This respect was intimate, involving the sharing of fruit with Buddha, who would distribute it among the rest of the spirits as if they were his staff.

Each of the stories in this chapter—from the spirits who haunt the land after civil wars, to spirits who murder a rat, to spirits who avenge wronged women—engage with sensorial encounters that cannot be easily explained by me or my interlocutors. In so doing, they also shape ideas of personhood that are not fixed but mutable. They enter the realm of the extraordinary senses, materialities that evoke mysteries. I contend with these mysteries as ways my interlocutors grappled with the inexplicability of violence. In this way, understanding nonhumans such as spirits meant understanding the violence of war. The humans I knew related to the spirits who were prone to violence with an uneasy kinship, similar to the ways in which they related to other humans in the minefield's postwar ecology.

I do not propose that relationality on its own can repair human-human relationships after war. Moral philosophers represent relationality, especially when used as a critique of hyper-individualization, as a dichotomous solution to the value of individualization. Judith Butler (1995) explains that building relations based on autonomy undermines "a moral mode of being that is bound up with others."[5] As I point out in the next chapter, to understand persons as being primarily relational confuses boundaries in ways that circumscribe concepts of an individual's consent. Human entanglements with spirits who are potentially violent and must be appeased render them part and parcel of relations but also place them in a hierarchy of power. To appease them mediates their potential violence and makes them beneficial. While this points to a fluidity between beings and a porousness of bodies, it does not dismiss hierarchies of power. That is, when premising a being as relational, it does not always follow that beings exist on equal playing fields in a postwar ecology just because of the diffuse characteristics of personhood and agency. Likewise, the rat can be loved by humans and even influence and alter the humans, but he will remain a "*littlest* one."

In this way, then, we can think through the ramifications for concepts of justice and retribution when it comes to mass atrocities like warfare. "Evidence" for example, gets destroyed when the dead must be quieted and allowed back into the reincarnation cycle. When I first began thinking through reincarnation, I interviewed the Venerable, who had entered the monastery to "make up for what was done" after the Khmer Rouge regime. The Venerable explained to me the principles of anatta—that is, the "no soul" or "no self" philosophy in Buddhism (for how this relates to death and power, see Davis 2015). I asked the Venerable, "If there is no soul, how can there be reincarnation?" In response, the Venerable presented a puzzle. "When you light a candle with another candle that is lit," he asked, "is it the same flame? That is samsara [reincarnation cycles]." In this case, reincarnation of souls/selves, as in a new flame ignited by another, offers ideas about how reincarnation is an expression of relationality, through a self's dependence on another self for existence, as if the self is being reborn anew. The relationships between victim and perpetrator then also become more fluid, demanding that the children of the victims of the Khmer Rouge light incense at Pol Pot's grave or pray to the "Khmer Rouge dead." Accountability becomes transformed under relational premises, too. Likewise, relationality mediated the conflicts between former enemies on the demining platoon, allowing them to transform both themselves and each other "to be kind," as we will see with the new loves in the following chapter.

5 *Metta* Means "I'm Sorry, You're Sorry"

I wanted to be left alone more than others expected. I learned to accept touch when I normally would not have; I was told to smile when I was sick or sad; and I was encouraged to laugh and joke even though I am not naturally performative. This was a new way of behaving for me. When I sought privacy, I would be corrected by many people.

While doing fieldwork at the prosthesis workshop in the K5 belt, I could not take walks on my own for my own safety from the shifting landmines. This meant that if someone was not available to drive me, then I had to stay in the workshop confines. But fieldwork pushed my personal boundaries in other ways. On one occasion, I was trapped in my tiled bedroom, sick with fever, vomiting, and having diarrhea. I threw up in a ziplock bag, sealing it and placing it on the floor. I had to take a medical house visit because my interlocutors wouldn't want to drive me to the city, over three hours away by car, when I was that ill. Because of my illness and because of the landmines, I couldn't walk to the village doctor's house. When the doctor came, he advised me to take some pills. While I crumpled my body into a fetal position to wait out the sickness, hot and sweaty, my friend, the pregnant wife of one of the prosthesis technicians, came to lie down beside me, spooning me. Her hand touched my hip bone. This was for comfort and company in her eyes, easing me into healing. I relented, letting my friend touch me in the heat as I sweated. I stared at the bright yellow square of sunlight in my room where the ziplock bag glistened.

It seems possible to imagine from this tactile encounter that something about bombs and living in minefields make people more reachable—despite their personal desires. In that moment with my pregnant friend, my agency felt diminished or at least part and parcel of another's agency. Touchability, in this case, was an important sensorial encounter to understand the materialities of sickness and injury. It was also, I was told by people working at the prosthesis workshop, a good way to hasten the healing processes of the human body. Importantly, consent was either presumed or unimportant.

In disability studies, feminist studies, and race studies, scholars have discussed the ways in which humans violate bodies that are not considered the default or the norm. A lesbian ciswoman's body in early sexuality studies in the West was declared to be either "butch" or "femme," and in the writings of those doctors who examined them in invasive ways, the butch lesbian's "untouchability" or lack of experience in being touched was admirable, since it was closer to the masculine gender (Munt 2003). But what perspective are we taking when we view such dichotomous ideas of consent? The perspective of the very same heteronormative doctors. Such dichotomous ideas of consent can also be read as limiting and embedded in cis-heteronormative codes (Nelson 2016). To consider all touch as invasive is to prioritize touch as penetrative and does not allow for the toucher to be altered by touch. Consider an Indigenous child in Australia who is held down by her mother and female relatives to take medicine when she normally would reject it, a disturbing scene in Elizabeth Povinelli's (2006) ethnography that considers the question, What does it mean when to be cared for is also to be held down? On the other hand, touch often *is* invasive, and those who exist lower on a cis-heteronormatively coded hierarchy *are* more touchable (or less, depending on the context). I want to keep both these aspects of touch in mind when capturing relationality and how it was expressed in the postwar ecology between humans and rats in a Cambodian minefield.

When human deminers discussed their relationships with rats, they described a process of learning to love them. This process, importantly, involved a presumption of shared emotions that did not consider the rat's agency as individualized from the human's agency. "At first," Ravi told me, "I thought of rats of pests, but now I think of them as my very best friends!" In Ravi's construction, the thoughts about the rats are what change them through whatever they have done to convince him.

The words for love that the deminers used were usually from a set of three—*sralanh, anet,* and especially *metta. Sralanh* is a word that most easily translated, for me, into "love." This kind of love can be used for cute things, romantic partners, and even to express strong tastes. The other two words gave me pause. *Anet* usually translated as "pity." When I asked what *metta* meant, I was consistently told, "*Metta* means *anet-sralanh.*" That is, *metta* means "pity-love." It was through this that I understood hierarchy as part of the relationality between humans and rats, but also as diffuse, that individuals were rendered as part and parcel of each other but with a sense of hierarchy. Part of this was also expressed in the movement and touches on a minefield.

Metta is a word that may sound familiar to people with a passing familiarity with Buddhism. Metta (compassion/loving kindness) is one of the four boundless states. The other three are the closely related karuna (compassion/sometimes also translated as loving kindness), motita (loving joy) and aphika (equanimity). Normally, English speakers translate *metta* as "loving kindness," or "compassion," and *metta* and *karuna* often become conflated, but either translation omits valences *metta* has for Khmer speakers, who usually define it as "pity-love," and its alternative translations "compassion" or even "mercy" are a better fit. As one of the four "immeasurables" of Buddhism, metta has also been called one of the "four sublime abodes" that are understood to be states of being (Obeyesekere 2002, 179, 388). They are immeasurable in that they cross the boundaries between all things; that is, all entities have the potential to achieve these states. Through meditation, these four boundless states allow practitioners to come closer to enlightenment (Nyanatiloka 1997).

Metta, I was told, can "turn a cruel person kind" and could also help to de-escalate situations of conflict. I link expressions of metta on the minefield with the shadow stories of ambiguous actors of violence. Pity-love is diffuse and belongs to everyone. The term itself can be seen as an expression of passive acceptance that fits into the Buddhist "ontological resonance" Alexander Laban Hinton (2005) describes—that is, the "under-standings shaped by [people's] distinct backgrounds and experiences" (25). Specifically, he says that Buddhism offers people "local understand-ings of anger management [that] provide a powerful alternative to vio-lence in Cambodia" (32). Julia Cassaniti's (2015) work about villagers in Thailand suggests that this kind of anger management has to do with what she calls an "intersubjective affect," where individuals are subject to the wills of others and vice versa. Scholars like Tanya Luhrmann (2011) and Cassaniti believe this indicates that Buddhists hold a radically differ-ent "theory of mind," from other religious communities, such as Christians in the United States.

In minefields, among villages and deminers alike, I found that relational understandings, touch, expressions of love, and, for the deminers, connec-tions to the rats helped to mediate potential violence. In this way, agency became redefined as mutual and accountability as shared—a reflection of the ambiguous agency of violence and the diffuse sense of ghostly persons in the postwar ecology where we worked. This inevitably involved, though, a transgressing of boundaries that sometimes dismissed mutual consent and, despite representing relationality, often reinforced hierarchies. The rats, for example, had not consented to being taken from the wild, bred according to their scent-detection talents, and forced to work during the day although they normally would be awake only at night. The introduc-tion of the rat was itself a transgression, and along with this transgression came a new choreography of the landmine-detection-rat technique.

One of the very first questions interlocutors had for me as a researcher was how I felt about animal rights. I had to reassure the landmine-detection-rat NGO workers that I did not work for animal rights organi-zations nor was I affiliated with them. "Good, good," Liz told me, "because we sometimes have problems with people from PETA."

Indeed, multiple responses to my work from fellow academics have voiced concerns about animal welfare. Sometimes this comes in the form of presumed nonreciprocity of the love for the rats. Typically, I hear, "Well, sure the humans loved the rats, but how do you know the rats loved the humans?"

Normally, in response, I say that I could see the ways in which the rats snuggled with the humans and followed them off leash. This often remains unconvincing to respondents in the academy. Part of how I understand the reluctance to believe the rats loved humans (besides an anthropocentric view of love) is by thinking through presumptions about reciprocal love: (1) that love can never involve hierarchy and (2) that it must be expressed independently by autonomous individuals.

Love, as a concept, *feels universal.* In fact, love feels untouchable by postmodern contingencies, because to suggest that people love differently might verge on unethical othering. But historicizing Western-centric ideas of love as requiring individuals who stand on equal footing with each other helps undo the assumptions that the rats and humans cannot really love each other—in the postwar ecology of the minefield, love for landmine-detection rats was both relational and hierarchical.

Humans who have been categorized as "abnormal" often have their agency diminished so that they also become more touchable—their individual personhood does not matter as much as the person who violates their bodily autonomy. When my pregnant friend cradled me, I was sick, a woman, and also, foreign. My body, notably, in a way that was similar to that of the pregnant woman who touched me, was touched in a way that seemed to suggest it belonged to others as well as myself. Most of the time, this is described as violence, which it is if transgressing individual autonomy is automatically violent. I do not want to insist that violation was never at play in some transgressions, especially considering the patriarchal and hierarchal norms of Cambodia, but I do want to consider that transgressions of individual consent, as when I was touched while sick, were part of the relational practice of the minefield's postwar ecology.

Anthropologists like Faye Ginsburg and Rayna Rapp have written extensively on how disability offers reconfigured categories, expanding ideas of kinship (Ginsburg and Rapp 2018, 185–201), healing (Ginsburg and Rapp 2013), and ways of being human (Ginsburg and Rapp 2020). They describe how disability in the United States, for example, "widen[s] the space of possibility in which relationships can be imagined and resources claimed," because vulnerability confronts people, rendering public intimacies that are normally private. For them, disability itself depends on relationality in its construction. In a minefield, part of the postwar ecology's daily relations were the visible disabilities of explosives' impacts on human bodies (Pardo Pedraza and Morales Fontanilla 2023).

Bombs have rendered everyone potentially vulnerable to injuries, and in a postwar ecology, amputations are a common sight. Cambodia today has one the highest percentages of amputees in the world, with over 40,000 amputees among its population of 16 million. Most amputees live among the mines, concentrated in the rural peripheries where the bombs continue to plague them and where accessibility is more difficult than in the cities. Their prevalence makes it apparent to all how vulnerable human bodies are to such terrible accidents.

When all transgressions become assessed as violations of varying severity, we lose the ambiguity of some touches. I did not want to be touched when I was sick, but I understood that my friend thought it would be helpful and I was grateful for her intentions. Thinking about transgressions beyond the binary categories of repulsion or acceptance helps me reframe the relationships I encountered in the field as revealing something about differences in understandings of agency and individual autonomy. To me, ambiguities about consent related to premises about what it means to be a person. What happens when individual autonomy is not a given or a primary value? As I have shown in previous chapters, agency and accountability for violence in the aftermath of Cambodia's war were depicted as ambiguous, and ghosts manifested a kind of diffuse personhood. Postwar ecologies materialized these relationalities in sensorial encounters that often took expression in Theravada Buddhism and animist cosmologies during my fieldwork among deminers and former combatants.

·   ·   ·   ·   ·

Simon the rat finished his detection work, and his handler, Moch, unclipped his harness and guided him to his travel crate. Moch patted her leg and the rat bounded after her—an action that delighted anyone watching. Simon was the biggest rat. His body was three feet long including a tail which took up half that length. His rump maneuvered through mowed grass to follow Moch's footsteps. Unlike the other rats who took some time to get used to the Cambodian handlers, Simon had always followed Moch.

"Little sister, little sister," Moch said to him.

When asked why she called the four-pound male rat "little sister," Moch answered: "*Sralanh* [love]. I don't know," she said. "I know he is male, but I just feel like he is my little sister. *Khnom snaeh* [I love (him)]."

Sralanh and *snaeh* are words in Khmer that indicate love that can be used for cute things, romantic partners, and to express strong tastes. *Snaeh* also indicates a kind of entangled attachment. Even though Moch felt love for Simon right away, she said that when she heard of landmine detection with rats and applied to be part of the first team to use landmine-detection rats in Cambodia in 2015, she "did not like animals. Especially not rats!"

As soon as she saw Simon, however, she loved him. For the first time in years of conducting ethnographic fieldwork on minefields, I heard expressions of love coming from people who detect explosives. Over the course of my ethnographic fieldwork with the implementation team for the landmine-detection-rat technique in Cambodia, I saw that the way the human landmine detectors (a.k.a. deminers) learned to love the rats also changed how they learned to live and work with each other. Deminers' labor in the minefield was inherently aspirational—"to help my country," they often said. In the minefields in Cambodia, the labor to decontaminate is nothing if not future-oriented. Rats have been employed into this future imaginary. They work with former soldiers who use their military expertise for this decontamination, which often produces tense workplaces where former enemy combatants must labor together. Postwar ecologies—that is, ecologies that can only ever be aspirational due to the endless nature of the problem of war and military waste (Reno 2020; Zani 2019)—rely on hope for the success of minefield clearance with rats and with enemies. According to their own definitions of the word, to profess *metta* allowed Cambodians who had once been cruel to become kind. This expression had a particular valence for deminers who had to work with former combatants but the stigma of past violence made it uncomfortable to directly address such former enmities. Loving the rats sublimated the minefield's potential violence into potential pity-love for and with colleagues. In a context where violence and guilt are not addressed directly, I ask how love for a rat might "make a cruel person kind"—that is to say, how rat love may mediate a violent past to work toward a previously unimaginable future.

The concept of metta as it was expressed to me in the field allowed postwar relations to become possible, and its entry point was in the multispecies relationality between humans and rats. This relationality spoke to a diffuse sense of personhood, a pathway toward postwar aspirations, and reframed lived-in presuppositions about individuality and kinship. Unlike love for animals as a moral obligation dependent on binary differentiations through an ethics of indifference (Dave 2022), detachment (Candea 2010), and desire or passion (Berlant 2012; Luhmann 1986), Moch's love for Simon allowed her to practice other loving relations within her demining team that diffused pity and love throughout its members. Rats like Simon thickened relationality among deminers.[1] This is not only a love story, though. This love depended on technologies of war. Rat-human relational-becoming accounted for the ways in which violence was already always shared, which made the recognition of kin networks beyond human beings crucial to mediating the minefield's dangers.

As Marilyn Strathern (2005) has suggested, personhood is constructed according to different conceptual foundations. She considers how personhood, based on individuality as expressed in the Western legal system, conflicts with various practices in places where the legal system gets imposed—for example, in Melanesia, where she describes personhood as constructed with relationality as its premise. In this case, a person cannot exist without relations to others, which implies different ideas about consent and agency. Sokhieng Au (2012), a historian of medicine in French colonial Cambodia, clarifies clashing systems of legality as "cultural insolubilities" that are caused by conflicting understandings about power and selves. Insolubilities, she says, are moments when conflicting networks, such as the Cambodian colonial subjects and the French colonists, were forced to interact with each other through economic, historical, or other kinds of necessity but could not be "smoothly integrated because they would lose internal coherence" (2).

In one such case during the colonial era, Cambodians beat a corrupt French tax collector. The king of Cambodia decreed that the village be fined and give merit every year on the date of the tax collector's murder. The king did not single out individual agents, and his penalty conflicted with the French colonial solution, which was to punish individual criminals in a trial (Chandler 1982). This case shows that Cambodian governance favored more distributed accountability, which clashed with individualism represented by the French manner of justice. In addition, the Cambodian king invoked Buddhist practice that called upon karmic connections between the living and the dead—that is, the perpetrators and the victim.

Au (2012) avoids the use of *incommensurability*, as that implies that people are not able to understand the logic of another. Rather, *insolubility* suggests that the logic and networks cannot accommodate each other or coexist even if they may be understandable to all the participants. In anthropology, the ontological turn has sometimes implied a sense of radical incommensurability between the ways in which people think and know the world. Philippe Descola (2013), for example, has described four different ontologies that categorize ways of ordering reality (animism, totemism, analogism, and naturalism). Such orders of thought condition possibilities of what it means to be human (and nonhuman) among the people who live within them. The crux of such a systematic catalogue of how people think is not actually its universalizing tendency but rather where people can mesh these orders and how people manage to think with each other. Some scholars of the ontological turn find these orders of reality as grounded in the logic of what *is*, combining evolutionary scientific evidence with modes of representation and thinking (Kohn 2013).

As a scholar of human-animal relations, I find ontologies useful ways to think with difference, but I also hesitate to draw from them, because my question always lands on how radical alterity really is—whether humans can truly know what it is like to be (an)other. The question itself comes from the philosopher Thomas Nagel (1974), who claims that human imaginaries could never know what it is like to be a bat. In a thought experiment, Nagel concludes that humans can never know, because being a human with human consciousness is so radically different from being a bat. I find myself unconvinced by his logic, which depends on what he calls "the mind-body problem," and I find that across ontologies, humans can know what it is like to be each other, so why not across species? Indeed, the very practice of categorization in ontological theory suggests this is so. In the case of my fieldwork, I prefer to use Au's term *insolubility* rather than considering differences ontologies, since she directly explains such systems as not incommensurable but rather as having practices that mismatch. How then, can I understand a system of practices where people do not consider themselves autonomous individuals but rather relational beings? And what does autonomy look like or become within such systems?

I reframe this issue of consent, which accounts not only for the touch between that woman and myself but also for other encounters of love and care in the minefield between the platoon of rat handlers and rats, as a mode of interacting that contends with relationality as a premise of that interaction, a different formation of autonomy. Normally, when scholars discuss relationality, it becomes a way to critique individual autonomy, leading to a sense of innocence about the ways in which relationality may provide ways to overcome disparities and a lack of empathy amidst global horrors.[2]

Other scholars of Southeast Asia have noted this radically different way of thinking about selves and others even when Buddhism is not present (Wikan 1990; Throop 2010). These scholars explore the disciplinary control that relationality of this kind wields over people as opposed to the radical alterity of a religious-oriented cognitive process. When a person believes that thoughts have actual effects in the world and over the wills of others and vice versa, it may enforce political or social control over behavior. Jason Throop (2010) points out that intersubjectivity in his fieldsite in Yap in Micronesia caused empathy to be a negatively valued trait, emphasizing a moral imperative for "self-governance" (772). This kind of understanding of selfhood takes into account disciplinary power, where understandings about humans as subjects and how they love are powerful ways to maintain "group cohesion" for nationalistic purposes (Brown 2005, 30).

But these divided theories—those that take on a more religious analysis of relational selfhood (Cassaniti 2015) and those that examine disciplinary means of power within relationality (Throop 2010)—are not mutually exclusive. In Cambodia, relational expressions of love can both be a form of control as well as a radically different way of thinking about selves. Perhaps relationality allows for a wholly different "psychic life" (Stevenson 2014) for people in Cambodia, as opposed to the psychological response to power that produces individualized agentive subjects more common to places like North America. Although relationality may allow violence to become something that needs to be accepted as harming everybody, producing a kind of passivity, it can also be used to do the work that, as Obeyesekere (2002) suggests, deals "with basic human problems" (xx).

On the minefield, the practices and expressions of love expanded boundaries of individual autonomy such that one could love a rat and that rat, despite being unequal, could reciprocate. Moreover, the ways in which the deminers expressed their love for the rats revealed that love itself is a pervasive essence that exists beyond individuals. This love transgressed boundaries even as the minefield itself insisted on materially enforcing boundaries and limitations.

.   .   .   .   .

Those first few weeks in Cambodia, Liz and Chann were still negotiating for electrical generators so that air conditioning for the rats would stay on even during the rainy season's rolling blackouts. The rats were still getting used to the newness of the place. During the evening, they went on leashed walks, sensing the new deminers' guiding touches through tugs of the strings on their harness. They were used to the feel of the harness but not the new smells or the new feel of the ground beneath their paws. They squeaked all night after finishing their nightly walks. The human handlers in the next room would toss and turn, hearing them through the glassless windows. The rats settled into sleep around dawn. Their kennels had an air conditioner but sometimes, in the power outages common during Cambodia's rainy season, the electricity went out and the kennels became unbearably hot even with no sunlight.

One part of the rats' routine remained unchanged from their time in Tanzania and Mozambique. Just after they quieted, when the red light of dawn was just a line over the horizon, a human entered their room. The light at that hour formed silhouettes of the temples in the far-off distance of the flat horizon. The rats didn't see that, though, since they are mostly blind. They curled up in their terra-cotta pots; the only one who continued to squeak during sleep was Merry.

Tokla, the vet, swung open Merry's wire-crate door. The first thing Merry did at the feel of the soft finger stroking his back was lift his head and point his nose at Tokla's hand. Tokla hovered it above the rat's nose. His squeaking stopped. Merry's eyes were only half open but his whiskers and nose were alert, twitching. "This is how he knows I am here," Tokla said, explaining the method to me later. "The rat smells me and then I can pick him up."

The next step, picking up the rat, required a gentle touch, too. Merry did not know Tokla very well and even though he was used to humans, his life had been suddenly altered. He did not bite Tokla's fingers, but he pressed his nose close against the skin. Tokla stroked Merry's back again and then lifted him up, following the instructions of Mohammed, a consultant from Tanzania, who had shown him how to do it the first week. The vet folded his own legs beneath him and continued to stroke the rat's back with one hand while the other hand readied a bottle of sunblock.

In Cambodia's wet season, the sun rises rapidly. The first step of land-mine detection with rats was to smear each rat's sensitive areas with sunblock. Tokla dabbed the white lotion on Merry's ears and then massaged it until the rat's skin absorbed it. He did the same on the two-toned tail, black near Merry's rear and changing midway to a pinkish tone. He put sunblock on Merry's paws last. Some of the rats handled this better than others. Merry was still sleepy so he relaxed in the human's hands, rolling to his side. Tokla dabbed the paws, and his touch was tender as he rubbed the lotion into the furless feet. Tokla repeated this process with each rat. By the end, all of them were awake from his smell and the sounds of him moving throughout the room. The sun gleamed white light, creeping around the border of their window's curtain. The rats recognized the sounds of crate doors being opened and closed. This was the start of their workday.

Touch was an important interaction between Tokla and the rats. Some of the deminers helped Tokla in the mornings. Moch, Ravi, and Sovannah would sometimes join him. The task was not popular because it resulted in the most nips from the rats. I would come to the minefield each morning after the dawn ritual, and Tokla would shake his head when he saw me. He would splay his fingers in front of me. "See? They bit me again."

Despite this frustration, the massages each rat received in the morning were essential for building trust between the humans and the rats. The humans understood the rats' vulnerability to the glare of the sun. It became a point of connection they had with the rats. As Moch told me, "You need to wear many layers of clothing because, in Cambodia, the sun cooks our skin." In the minefield she wore turtlenecks and fingerless gloves as well as a hat with a neck flap to protect her. While for her, this was about retaining a pale skin color as well as health, she understood the rats' need for protection from the sun, too, which is why she volunteered to help Tokla. The sunblock also helped illuminate rat behavior—every deminer knew that the rats had not developed protection from the sun because they were nocturnal. It also helped them understand how sluggish the rats could be in the later hours of the morning because, the deminers told me, it was their "sleepy time."

Meanwhile, through this touching routine, the rats familiarized themselves with new humans. Notably, the rats had the most "proper behavior" with the deminers who regularly applied their sunblock. "Proper behavior" meant that they followed their training and were more habituated to the humans. They attended more to the pull of the strings by the deminers who touched them each morning. They also, eventually, bit them less. Moreover, the morning sunblock routine in Cambodia mimicked the routines in their original homes. Mozambique, Tanzania, and Cambodia are all close to the equator, which means exposed skin has to be protected. The similar patterns helped the rats to associate daybreak with the beginning of their workday, detecting the scent of TNT beneath the soil.

These vulnerabilities were integral to a mutual understanding between the rats and the humans. Partly this was about exposing mutual vulnerabilities—the rats discovered they could pierce human skin, and the humans discovered how fragile rats were to strong sunlight. The rats' skin sensitivity became tragically apparent when one of the deminers felt a bump on Marcous, a particularly heavy rat. The bump at the edge of his ear was the beginning of skin cancer. Marcous ended up being retired shortly after APOPO gained certification from the government. The handlers who had massaged sunblock on his ears were the ones who cried the most when Marcous died. This touch opened new ways of relating and new ways for concepts of Buddhist love to find purchase on the minefield.

·   ·   ·   ·   ·

After I had hung out with the deminers for months, we became very close. I would follow them during their work in the morning, help make food, and then, in the afternoon, attend the courses on "theory," which were courses on landmine-detection technologies and explosives buried in the landscape. On some days, after requests by the deminers, I was asked to provide English-language lessons, which I did for the rest of my fieldwork stay.

As I showed with the story of my getting sick, the setting of a minefield seemed to alter the very formal touching etiquette in Southeast Asia. Physical taboos about touch do often align with social hierarchies (Aulino 2019). For example, women should never touch monks, and no one should touch the head of a person—the head corresponds to the part of the body where spirits and can enter and leave. But these taboos also correspond to politics and to ideas of a society being interdependent in all its parts, as well as being hierarchical. In contrast, the atmosphere on the minefield was characterized by teasing and joking touches that disrupted these taboos. My only equivalent of such an atmosphere was the stories I had heard from my father about his work as a firefighter—he would often talk about the bawdy jokes and pranks the firefighters performed on each other. I imagined that the interaction among the platoon of deminers was similar to the comradery of a barracks.

Workplaces like this have been represented as a toxic masculine environment and fairly unwelcoming to women and nonbinary folk (Anjum et al. 2018). I did find some of the jokes a bit overwhelming and sometimes even physically invasive. This was yet another moment when touch taboos were overcome on the minefield. My body, as a feminized and foreign body, was overtly sexualized in many of these jokes, and I relied on the two women and the nonbinary person to help me navigate some of them, both by mimicking their responses and playing along with the jokes.

For example, once, while I was filming Meas washing the rat kennels, a deminer laughed and said, "Ah, beautiful brother, the white woman is really interested in you!" Meas's response was muted, and they wore a set of headphones—even though they smiled at the joke, they did not stop their work or play along. When their male coworkers talked to them, the men would call them "beautiful brother," which was a reference to their third gender in a mocking way. The women would just call Meas by their honorific, which was more common—that is, *bong* (older brother/sister) or *ouen* (younger brother/sister). Khmer conversation provides an easy mode to communicate nonbinary interpellation because normally the brother/sister part goes unsaid. Eventually, the men stopped and just used the honorific after Meas ignored them long enough.

The other strategy was one employed by Moch. At one point, Ravi made a joke about me—not one that was overtly sexual but just about my appearance, and Moch said, "Oh, Darcie, you should kill him!" So, following along with what I had seen her do, I reached over and tapped him lightly on the shoulder, in a pretend slap. The whole car laughed but Ravi seemed shocked. I had learned how to touch differently on the minefield with my interlocutors.

After that, Ravi and I exchanged more jokes with each other. He and I spent a lot of time talking about how much we loved the rats. When I returned for follow-up field visits, I was sure to visit his platoon. Unlike those of some of the other platoon members, his jokes did not involve a degree of boundary pushing that felt uncomfortable to me, but they did represent the way transgressions allowed the platoon members to relate to each other. Living together in two rooms, sharing such a small space, and learning to love rats made for eventual bonding that overcame potential fears in the platoon. Transgressing some of the boundaries helped to mediate some of the fears.

Love for other humans could go unspoken in the minefield. Like the fears portrayed only by shadow stories, when one of the handlers loved another handler, these emotional expressions were communicated indirectly. Love for the rats, on the other hand, was very direct and vocal. These direct expressions of love for the rats became ways to indirectly express love for humans, which, indeed, mediated the shadow stories of fear.

I am interested not only in depicting how rats relate to humans and vice versa but also in the relationships humans have to these relationships. Rats and humans emerged as new kin (one woman called her rat "little sister"; others said rats were their "very best friends"). These new relationships were important because they improved the confidence of the humans in the rats and in each other. These relationships unsettled "reductive categorizations" (Hinton 2016), such as the polarizations of perpetrator and victim, which were as much remnants of war as the landmines were.

Love, while crucial to landmine detection for building the platoon as a unified team, was also important because it actively disrupted existing categories. Not only was the emotional expression of love important, but how this love operated, the ways in which it was voiced. In the minefield, the deminers would use words of affection for the rats, *anet* (pity), *sralanh* (love), *snaeh* (love, sometimes defined as entanglement or romantic love) and *metta*, which they consistently defined as "pity-love." Of these words, *metta* seemed the most out of place because it is not only a word for emotions but also a religious idea. By analyzing *metta*, we can understand why the way the humans related to the relationship was as important as the relationship itself.

People would often translate *pity* as a reciprocal feeling directly into English even if they were fluent in both languages. For example, in 2010, a Khmer man professed his love for me in English. Even though I barely knew this man, he told me that when he saw me, "he felt pity for me" and he wanted to express his "deepest love and pitifulness" to me. Note here that he expressed pity as both "for" me and "to" me, as in his own "pitifulness." For him, when he pitied me, this corresponded with pity from me. At the time, I felt worried by this interaction because of its boundary crossing and decided to distance myself from this man. But other Cambodian people who proclaimed how they loved me would often simultaneously proclaim pity as well. When a demining supervisor tried to explain my vegetarianism to the other deminers, the deminers asked me why. I said it was because I loved animals (*sralanh*) but the supervisor gently corrected me, telling me that I felt "*anet*," or pity, for animals.

My confusion about pity mixed with love and about how pity could be reciprocal came, I think, from the connotations *pity* and *love* have in my own language. That is, pity is most often associated with power dynamics and pathos. Love is almost diametrically opposed to pity in this sense—it could never be associated with hierarchy because love itself is supposed to dismantle power dynamics. Or rather, as Wendy Brown (2005) has pointed out, "the lover finds her- or himself in a condition in which she or he must refuse all evidence of flaws in the object" (30). Pity, for me, indicates an object that is quite flawed in that it is weak and deserves pity. But perhaps the idea that being pitiful is being flawed is not universal. Moreover, perhaps the idea that pity and love automatically produce objects is not universal.

More and more in the field, I saw that metta came up in ways that differed from the hierarchical notions that pity would suggest. Or, as my friend Narith put it, "I feel more comfortable saying 'I love you' in English because it's not as big a deal. When you say any word for love in Khmer, it, well, first of all it's not normal because people do not say it as much [as English speakers], and second, it means something different." *Love* for him, in Khmer, was both more intimate and more powerful than in English.

Pity, as Felicity Aulino (2012) has pointed out in her anthropological study on Thailand's changing elder care services, emphasizes hierarchy in English, not intimate love. But this differs from how *metta* or even *pity* operates in her fieldsite. Aulino describes how *metta* at an eldercare facility where she worked seemed to reinforce a karmic hierarchy despite the intimacy between caregivers and those to whom they offer care. She describes her shock when a caregiver translated *metta* as "pity" because for Aulino, *pity* transforms love into something that is different from *love*. She notes, as an explanation of this shock, that even "the term pity (*song-san*) itself cannot necessarily be taken as an exact translation of the English" (2012, 110).

It is this moment in Aulino's ethnography that I relate to so well. *Metta* has confused me, too, because pity does not seem to be part of love. But to translate words is a risky business. Even in English, words, especially words like *love*, have changed throughout history. In his conceptual history of the Renaissance, Niklas Luhmann (1986) explores "a semantics of love" because while the word *love* may be the same word, "meaning changes . . . [and words] encapsulate particular experiences and open up new perspectives" (8). Luhmann suggests here that the multiplicity of potential meanings within words like *love* have power, even spurring on changes in the ways people think, which corresponds to connotations of *love* in Khmer. Rather than translating *metta* to "pity" or "love," we can see from behaviors, expressions, and stories, that metta may be a different experience from pity or love, particularly in the Cambodian post-conflict context.

As I have mentioned in the previous chapters, post-conflict contexts are multiple and have numerous ways of mediating "intimate enemies," but scholars have pointed out that reconciliation often entails an emphasis not on blame, but on religious ideas about "mercy" that are embedded in a kind of intersubjective framework similar to metta (Noordegraaf 2018).

That said, what I want to point out here is that from my work in Cambodia, post-conflict mediation entails intersubjective accountability, which also works as a kind of disciplinary power.

"We have metta for the rats," Ravi told me once when I asked him to describe how he felt about the rats. "It means that if they are sick, we will take care of them—we feed them, make sure they have everything they need. At first, I didn't like rats because I thought of them as pests. But now, I think of them as my very best friend."

When I asked questions about metta, I was often told to "go ask a monk." Indeed, I followed this advice. The monk I asked told me a fable that mirrored relationships I saw developing between deminers and rats:

> A pregnant woman is walking into a forest. And she has so much metta for her unborn child, that even when poisonous snakes, dangerous animals, or cruel men see her, they will have metta for her. It will mean that poisonous or dangerous animals will not harm her. This metta can turn a cruel person kind. Metta exudes energy.

The process of metta the monk described is a love that emanates from the object of affection; it emerges from the mother and child and then also affects the predatory animals. While doctrinal Buddhism as described by the monk may contrast with Buddhist practice on the ground, anthropologists have pointed out that in Thailand and in Cambodia these philosophical concepts carry over into everyday practice (Cassaniti 2015; Klima 2009; Ledgerwood 1994; Nam 2011). This love comes from the woman and the child together—her love for her child then becomes the love the onlookers have for her. This diffuse love is similar to the kind of metta Ravi describes, in that metta is primarily for her unborn child but also involves the animal onlookers. This speaks to metta's transformative effects that formulate connections between the woman, her unborn child, and the predators watching. It also speaks to a concept of the human embedded in a different genealogy than that based on European humanism where *human* is conceptualized as separate from *nature* or even where being human is conceptualized as being separate from other individuals who are also human (Fuentes 2010; Broglio 2013; Ruddick 2015).

According to the monk's story, animals, not only humans, can have metta. Ashley Thompson (2004) points out that the conceptual history of what a human is in Cambodia differs from that of European humanism, even if analyses of these concepts must integrate the influences of war and French colonialism. While there exists a hierarchy in these differentiations between humans, animals, and other living things, Thompson claims that Cambodian Buddhists conceptualize a human as part of a general connectivity to all living creatures. In the ontological turn, Descola (2013) describes an "animist-analogist" framework, where understanding the concept of being a self means being among and connected to all other living beings (see also Århem 2015). If we understand metta for and with rats, it makes it even more important to attend to the rats when it comes to understanding relationships postwar. It behooved me then to integrate human-animal stories not as separate and singular from the postwar ecology, but, in fact, affected by it. Love, too, became (re)configured in the postwar ecology.

The pregnant woman has metta and then this metta affects predators in the forest and makes them not harm her. The pregnant woman's metta for her unborn child makes animals and other cruel creatures metta her. Here the child, like the rat, works as a mediator for a pity-love that makes cruel animals kind. Similarly, when handlers said they felt metta for the rats, the rats were *mediators* in the way that Bruno Latour (2005) describes, as actants who determine relationships in the minefield through their own categorical and behavioral aspects (39–40). Ratlike attributes and adaptations to Cambodia shaped the relationships in the minefield.

I am also interested in the ways this love was expressed verbally or rather, in how the radical pro-drop language of Khmer offered opportunities for this metta to illuminate its diffusion. What remained unspoken and unheard in these expressions of love were the pronouns. To understand pity-love in the ways Cambodians expressed it to me, I had to take silent subjects and objects seriously. In other words, I had to think with Khmer speech when it did not differentiate between subject pronouns and object pronouns, especially with words like *metta*. Such reciprocity in pity-love transforms the relationship between subject and object perspectives.

Central, then, to my understanding of *metta* in Khmer is that peculiarity of its grammar: the tendency for Khmer speakers to drop pronouns. "Pro-drop" languages, as I mentioned when discussing shadow stories, do not clarify pronouns; instead, they often have verbal agreement patterns to indicate the subjects. Khmer is considered a "radical pro-drop" language—that is, it has no verb agreement to indicate who the subject of a sentence is (Simpson 2005). Although I am not claiming, as the Sapir-Whorf hypothesis would, that the grammar of the speaker limits conceptual categories of the speaker, I am suggesting here that the grammatical structure of a radical pronoun-drop language allows for a representation that resonates with traversing the relations and emotional expressions across species in the minefield. So, too, did the Buddhist concept of metta, a relational love.

·   ·   ·   ·   ·

Despite having been formally introduced during that initial screening of the candidates, deminers and rats did not have an easy relationship in the beginning. On the human side, the rats were part of a family of animals considered pests or food. The humans who worked most closely with the rats, all from the Cambodian countryside, had to learn from the Tanzanian consultants that these rats were not the same creatures as those carrying disease from the sewers into people's homes, nor the same creatures as those that Cambodian farmers hunted and ate.

On the nonhuman side of this encounter, the rats had to recover from their long plane ride to Cambodia and relearn the scents of the new soil and then how to distinguish TNT from that soil. The sun was hotter in Cambodia, and even the bananas tasted different. Most of the rats refused to eat Cambodia's most common banana, which led Hien, the Cambodian demining platoon leader, to buy the more expensive bananas. "The rats do not want the cheap bananas," he said. "They like these ones, which are called 'lady fingers,' because they are so sweet!"

The joke here was that the rats preferred women's fingers because they found women more beautiful and sweet. The ways in which the rats' preferences were sexualized were consistent subjects of jokes on the minefield. For example, after a rainfall at night, the rats walked through dewy grass and the water made them pause in the middle of their traverse across the minefield pit and lick themselves. Inevitably, the handlers, watching a rat stop and lick the dew off her legs or tail, would say, "You are already so beautiful. No need to take a shower like a lady before her date."

While the rats were often the butt of jokes by men in the demining unit, who compared their grooming habits to the habits of a woman getting ready for a date, all the deminers called the rats affectionate terms like "littlest child" or "little sister,"[3] even a giant male rat. Eventually, everyone, even the men, caught onto these terms of affection.

Sometimes, in the case of the lady fingers, the rats were characterized as lusting after women. The other more common way in which the Cambodians discussed rat love was through the humans' sexual attraction to the rats. At weekly meetings, people would assess how work with the rats was going. When supervisors asked if the deminers had been bitten or scratched by the rats, one deminer said, "It's okay. They are only little scratches. It's the same as the scratches from a woman." Here the deminer was referring to scratches made by women in the midst of orgasm. His joke, likening the rat to a woman to whom he gives pleasure, made some people at the meeting laugh uproariously and some smile only slightly.[4] One interpretation could allow for a very gendered consideration of the rats here, as human women are also objectified in Cambodia, but, given how these jokes came alongside professions of love for the rats, it is also interesting to underline how affection and potential love was expressed by the comparisons to lovers, sisters, and friends.

The rats themselves did not usually respond to the sound of their names despite having been given those markers of individual personhood. They followed the gentle tap of a human's hand or quiet kissy noises rather than a name. Their favorite incentive was a snack offered at the end of a long walk across a mine-filled pit, rewarded once they successfully communicated that they had detected a bomb.

It is difficult to assess whether the love the human deminers feel and express for the rats was reciprocated in the way humans understand it. Certainly, rats do not consider pro-drop as insightful into the ways humans could actively disperse their love beyond a single pair of beings. Nor have they been trained in Buddhist philosophy (in this reincarnation samsara cycle, at least). The rats cuddled. They provided nonhuman modes of representation to communicate their calm and acceptance of their humans (Kohn 2013). But the rats also had a difficult time adjusting to the new scents and tastes of the country to which they had been exported. Simon was called "the best rat" not only because he usually indicated the correct location of a landmine but also because he was the most affectionate rat and never bit any of the humans. Once the rats were more habituated to the new setting, more of them showed signs of reciprocal affection to the new humans.

Victoria, the rat, began her tenure in Southeast Asia by running away from the Cambodian handlers. This sometimes led to panicky chases around the kennels or even in the minefield as she zipped around while the humans would crouch to try to grasp her and put her back into her crate. A few months later, Victoria and Ravi, one of her human handlers, would be seen regularly snuggling together. When Ravi held her, she would crawl up his arm to lick his neck. These were "rat kisses," and they showed the deminers that the rats, even the stubborn Victoria, loved the humans back.

·   ·   ·   ·   ·

As noted, the deminers did not separate sleeping quarters by gender, instead dividing them along institutional lines, but the women and the nonbinary person on the team of CMAC deminers did put up jewel-toned room dividers to keep their cots separate from those of the two men with whom they shared a room. The rooms were also where the deminers distributed the luxuries their organizations had given them: the CMAC room had the television set, and the APOPO room had the fridge and coffee maker—an electric kettle to boil water that was poured over water-soluble coffee crystals.

I would often relax with the deminers during lunch break in the CMAC room. The APOPO deminers would hang out there, too, because the TV set had bigger appeal. The APOPO deminers would bring coffee and cookies for snacks before lunch was made.

After eating a lunch of bitter herb soup with spiky veggies, Moch and I, as usual, lay down on Sovannah's cot. Moch said it was perfect because it had a good view of the TV. Instead of watching the historical-drama soap opera that was on, we giggled together as she showed me new photos of her brother on Facebook.

We were in the CMAC deminers' bedroom. The CMAC deminers had drawn a chessboard on one of the floor tiles and played a Khmer version of chess. Sovannah's cot also had a full view of the window, and the sun shone on us as we lazed about.

"Did you know, Darcie," Hien, a supervisor, said, interrupting us, "that we have a real love story here? Sovannah and Moch love each other."

I looked at Moch and she shrugged. Sovannah avoided my gaze. She seemed embarrassed at Hien's loud claims so I just whispered to her, "Do you love him?"

"Not really, I like him like I like anyone else," Moch said. She took a sip from Sovannah's water bottle, and I sensed then an unspoken intimacy in that sip. I realized that we might be lying on his bed not just because it had a good view of the TV or because it was nice to lie in the sun. She then diverted the conversation away from Hien's mention of love.

Hien claimed that Moch and Sovannah loved each other, but they did not directly state this about each other. There were small things that indicated their affection—when Moch and I were in the room together, she always guided me to Sovannah's cot and she drank from his water bottle. Most often, Sovannah chose to do tasks that helped Moch's tasks. As she cut a fish for dinner, he ripped up some herbs. These subtle clues were unspoken, in contrast to the open ways in which both Moch and Sovannah expressed their love for the rat Simon.

Simon was Moch's only rat after her other rat, Håvard, died. She and Samnang, an NGO worker, partnered to work with Simon. But it was not Håvard's death that made Simon special to Moch. He had been special to her since he arrived. "He is the best rat," she told me. "Proper and smart."

I watched Moch as she fastened the plastic buckles on Simon's harness. Its pink straps made his hair smooth against his skin, showing how large he was. He was the biggest rat and the best landmine detector. He weighed two kilos. Moch said she loved him.

When the rats finished detection work in their practice pits, the handlers unclipped their harnesses and guided them to their transport cages. The handlers patted their legs, and the rats bounded after them—an action that delighted anyone watching, though it took some time for the rats to follow the Cambodian staff as opposed to the more experienced trainers. Simon was the exception, because he had always followed his handler, Moch.

Moch patted her leg and encouraged Simon to follow her into his transport cage. "Little sister, little sister," said Moch.

Moch's love for Simon, like a mother's metta for her unborn child, can offer Moch protection. After all, these rats are going into the minefield before their handlers, ensuring that the potential explosives are detected before the humans can step on the unsafe ground. Indeed, like spirits, love for the rats took on diffuse properties—it spread toward everyone and created a kind of shared love.

Everyone smiled at one another when their coworkers succeeded in getting a rat to follow them. I remember Meas, very proud of their rat, Victoria, saying, "[She] loves me," when Moch and I watched the rat follow them. People agreed with their assessment and repeated their phrase: "Yes, [she] loves [you]!" The words for love echoed throughout the minefield as people repeated them.

But Moch's love for Simon was not only for Simon. Moch and Sovannah were not openly affectionate with each other, but they were affectionate with rats alongside them. The rats carried or "translated" the affections between deminers; they offered an indirect way for humans to express connections with each other, materialized in the rat's walk across a minefield that connects two deminers via rope (Kockelman 2010).

For example, while Sovannah and Moch did not openly hold hands or touch each other, Sovannah would pick up Moch's rat, Simon. Because Moch loved Simon, she would snuggle with Simon while Sovannah held him. At one point, Moch even kissed the rat on his nose in Sovannah's arms, brushing her cheek against Sovannah's arms. Simon became a way for the two to transgress the touch taboos of two young lovers on the minefield.

·   ·   ·   ·   ·

I also loved the rats. I especially liked Frederic with his need for nesting but I also really enjoyed Merry and his squeaks. Both aspects made the two of them seem vulnerable to me—Frederic who didn't understand that he would never be able to secure the walls of his nest with straw and Merry whose active dreams made him squeak. After talking about this with the deminers, we realized together that what made us feel affection for the rats was their vulnerability.

Indeed, as we have seen, the deminers used particular words for love that emphasized vulnerability of the beloved. The deminers would use the words interchangeably when we talked with each other about how we felt about the rats. This would sidetrack me. I would ask people to describe the differences between these words: *sralanh* (love), *anet* (pity), and *metta* (pity-love).

In the car, on the way to the market for a daily trip to get ingredients for lunch, I squished between Moch, Bunnary, and Ravi. Chann and Hien were in the front driving. This would be one of the few trips I was allowed to take to the market. I hindered the bargaining power Moch had in the marketplace. Moch told me that the vendors didn't want to give them good bargains on the food because I was white and foreign and they assumed that I was giving them lots of money and that the platoon was rich. On that ride, we were talking about how we loved the rats, and someone used the word *metta* and another person agreed, saying they had *anet*, too.

When Ravi had described how he felt about the rats, he said he had three words for his feelings: *pity, love,* and *metta.* "When a person feels metta for someone, they may want to kill them, but then, they feel metta and they do not anymore."

These loves had a particular resonance on a minefield, where the platoon members had tense relationships based on suspicions of mercurial loyalties. One supervisor constantly mentioned the presence of spies on the minefield. The deminers who worked under him, he said, must be spying for the government. He also consistently tied the government to stories of his experiences escaping the Khmer Rouge regime. Other deminers described how they had fought in battle. Separately, they told me which "parties" they had been on, and through this I was able to determine that several had fought against each other. At first, they seemed to mistrust each other, with rumors of espionage and frustration over pay circulating between the NGO group and the government group. This mistrust contrasted with the love and affection they expressed for the rats. Love, as Naisargi Dave (2022) points out in her work on animal ethics in India, is entangled with ideas of justice that are inherently violent. Loving animals especially relies on a hierarchy of indifference that does not allow hope for their future (Dave 2018). But the deminers' love for the rats meant that they could work within a *potential postwar* ecology that countered the dangers underfoot. Loving rats in Cambodian minefields allowed former enemies to love each other by de-individuating personhood even across species. These stories of rat and human landmine detection avoid what Dave (2022) calls multispecies anthropology's performance of loving animals like rats to explicate violent structures and hierarchies (657). Instead, living and working toward a postwar ecology required landmine-detection rats to be integrated into aspirations, through a love that transformed relations between enemies.

My questions about metta finally caused Hien to switch to English for me.

"Metta is easy to understand. *Metta* means 'I'm sorry, you're sorry.'" As he said this, he spread a hand in the air between us—indicating himself with the *I* and me with the *you*. To his dismay, this provoked further questions from me.

"You mean at the same time?"

"Yes, yes. I'm sorry, you're sorry. I forgive you, you forgive me."

Hien conflated subject and object nouns explicitly with *metta* by saying: "*Metta* means I'm sorry, you're sorry." His repetition of *sorry* with *I* and *you* suggests the interchangeability of *I* and *you*. Moreover, he conflated an apology with forgiveness and connected this with love. This conflation and reciprocity of subject-objects, specifically handler and rat, plays out in the interdependency rats and humans have on the minefield. *Metta* described a rat-human entanglement that made the actual interdependency in their work even more explicit. Through thinking with metta, hierarchy becomes part of the relationality between humans and rats, but personhood becomes diffuse, so that individuals are rendered part and parcel of each other. This thickening of relationality, this becoming-relational, was expressed in the language, the movement, and the touches on a minefield with the arrival of the rats. Moreover, perhaps the idea that pity and love automatically produce objects is not universal. Rather than translating *metta* as "pity" or "love," we can see from behaviors, expressions, and stories that *metta* may be a different experience from pity or love particularly related to the Cambodian post-conflict context. The difference lies in how it expresses becoming relational with landmine-detection rats.

·   ·   ·   ·   ·

There is no denying the practical interdependency between rats and human deminers. The human deminers depend on the rats to keep them alive—that is, safe in a minefield. The rats are the front line, detective agents who smell TNT that the humans cannot sense. The rats depend on humans to feed them and care for them, and as one deminer explained, "We have metta for the rats. We take them to the hospital when they are sick." In the ways that the deminers defined reciprocal love and pity, both humans and rats had love and pity for each other. Deminers told me that their mine-detection rat was their "very special friend" because she was so good at keeping them safe in the minefield. With the exception of Moch and Simon, it was only after training with them, learning mutual vulner- abilities, and working with former enemy combatants on a minefield that rats became their handlers' "very best friends," "little sisters," "dear cute ones," "littlest ones," and so on. Like the Hindu-nationalist cow lovers in Radhika Govindrajan's ethnography (2018), who nonetheless consider acts of labor part of cultivating love for rightwing means in India, the deminers felt metta for the rats after they labored together. In critiques of employing animals to work, anthropologists have pointed out that nonhu- mans are employed in work that supports global capitalist systems, such as the primate sanctuaries Juno Salazar Parreñas (2018) encountered in Malaysia, where postcolonial legacies employ primates to playact a decolo- nized imaginary.

The difference for the rats in Cambodia's postwar ecology is the mutu- ality of this labor, the interdependence of the rat-human-human configu- ration. The rat detects the bombs, which are dangerous only to the humans, not to the rat. The humans feed and protect the health of the rats. Terms of affection for the rats also reconfigured humans' relation- ships with their fellow humans such that the relations co-constituted the rat-human-human team. Vinciane Despret (2004) describes a similar process of co-constitution as "an 'anthropo-zoo-genetic practice,' a prac- tice that constructs animal and human" (122). Despret links trust to love and suggests that trust is a matter of expectations that are met within rela- tions for both animal and human (114). The rats, when they detected landmines in Cambodia, had to overcome mistrustful expectations that existed before their arrival. Loving the rats meant that trust could be a new expectation.

In this way, co-constitutions are about becoming responsive to expecta- tions. Such a trust is a new way of "becoming" a rat—what Despret (2004) calls "being-with-a-human" (122). *Becoming* (instead of *being*) provides a way to describe the potentiality and uncertainty of the relations they entered. As I pointed out in the first chapter, the design of landmine detection with rats produced a becoming, a being-with-a-human, or more accurately, being-with-two-humans or perhaps being-a-human-with-a-human-and-a-rat. But this being is more accurately describing as a *becoming* inseparable from the relations around it, which were destabilized by ambiguity and uncertainty. The possibility for ratlike love to emerge had to do with a future temporality. The rats and humans were *becoming relational* amidst and in response to the potential violence that surrounded them.

In the case of Simon, the rat acted as a mediator of affection between Moch and Sovannah. In other cases, the materialities of the rat-detection technique lent themselves to countering the suspicions the coworkers had about each other. Being tied together by string and forced to face each other and step in unison concretized the metta and other forms of love on the minefield. The sensorial encounters of twine connecting two deminers made me think of imagistic parallels with cosmologies of Theravada Buddhist *sorsai*. When describing the spiritual connections between all living things, Cambodian Buddhists use the word *sorsai* (សសៃ), which means "thread, filament, string," and also "veins." When people light incense sticks, they offer homage to the veins that run along and through the world's living things. One of my favorite artists, a contemporary artist from Cambodia, Sokuntevy Oeur, depicts these *sorsai* in her series of paintings called *Human Nature*. I take my understanding from testimonies of deminers who describe the *sorsai* to me as well as from her visuals. In one painting, a woman lies in a boat with a bouquet of flowers in place of a head, with red and blue veins squirming from another woman's mouth and hands to three tropical birds in the sky above an ocean (figure 8). Oeur's paintings consider this interdependency between humans and nonhumans as far from straightforward. While hybrids and the figures she portrays may be expressing love, they also hold violence within their relations (Oeur 2018; Lê 2021).

I also like to put *sorsai* in conversation with "string theory" from Donna Haraway's thoughts on how ideas and materialities interplay with each other (1994). String between deminers has resonance in Cambodian

Figure 8. A painting from the series *Human Nature* by Sokuntevy Oeur. Reprinted with her permission. Acrylic on canvas, 2014.

cosmology. It is this vein that Oeur, the artist, represents when she paints human-nonhuman hybrids wriggling from figures and technologies alike. This is the same vein that Cambodian parents represent when they tie string around their children's wrists to keep them healthy and safe. If this is a vein, then entanglement is not the tying together of separate entities but a place where blood passes through, a place where every living thing is revealed as simply one pulsating being. What if we imagine this string as something that extends our bodies so that they touch dangerous ground and connect with dangerous coworkers? Such an imaginary positions the rats, the humans, but also the minefield itself as part of one being, entwining violence with love.

It was not only twine but also Cambodian words that expressed emotions that tied these beings to each other. The deminers spoke of having metta (compassion or pity-love, as translated by Cambodians), sralanh (love), snaeh (entanglement or romantic love), and anet (pity) for the rats. These words implied transformations of the rat and the person who loved one another and the people around them.

During fieldwork, personal histories served as barriers to deminers' working together, barriers that needed to be overcome if the deminers were to depend on each other in such a dangerous profession. Metta was powerful. The love for the rats collapsed the binary between lover and beloved, revealing the interdependence of both. The rat kept the humans safe in the minefield while the humans "took care of the rats," which is

Figure 9. Moch, Sovannah, and Simon. Photo by Anonymous, 2016.

part of how they defined their love for them: "If the rat is sick or hungry, I will take care of them." "I will bring them to the hospital, I will bring them their medicine. I will feed them. And the rat keeps me safe in the mine-field. They are very talented."

Love for the rats and the jokes about the rats undermined a simple interpretation of the gendering of the rats as female and therefore sexualized objects. In fact, the rats allowed deminers to queer genders even beyond their jokes about having sex or romancing the rats, male or female. Moch, for example, called Simon, a huge male rat, her "little sister." Her gendering here of Simon's male body was deliberate, as she told me, "I know he is male, but I just feel like he is my little sister." Her feeling about his feminine relationship to her deconstructed gender as separate from his physical body, understanding gender itself as something beyond an individual essence and more relational, more mutable, like the diffuse spirits and the ambiguous actors of past violence.

The love story between Sovannah and Moch culminated in a marriage while I was away from the field (figure 9). Once I returned home, Moch got her own Facebook profile to connect with me. Eventually, the two posted photographs of their marriage. Under a photograph of Moch with Simon on her shoulder nuzzling her neck, Sovannah wrote in Khmer, "Beautiful sisters."

The process of overcoming former human enmity was related to the process of learning to love rats. Enemies became colleagues during the mine-detection training. The rats, formerly considered pests, underwent a similar transformation, from pests to deminers. These categorical shifts allowed humans to counter fears they had about their former enemies in ways that did not resolve conflicts but sidestepped them. While polarizations like friend/enemy, pest/companion, victim/perpetrator, animal/technology existed, they were traversed through forms of entanglement that blurred these categories, one such form being the process of mine detection with rats. Moving beyond these divisions, or rather, sidestepping them, gave the deminers alternatives to traditional models of justice and reconciliation. Like most relationships, the love took time to build and building it was not an easy process, but its expression countered the shadow stories on the minefield.

.   .   .   .   .

After war, perpetrator and victim are a pair that suggests individualized subjecthood, where one person is accountable for their actions. While this individualized subjecthood does not correspond to Theravada Buddhist theories of mind or Theravada Buddhist ideals about reconciliation, as Alexander Laban Hinton (2016) points out, part of the Khmer Rouge regime was an "either-or" framing of Khmer revolutionary or anti-revolutionary. By contrast, metta offers alternative framings of mutual accountability. I encountered this in the field. One of the deminers who had suffered in the work camps under the Khmer Rouge regime lit incense at Pol Pot's grave. When I visited the holy Preah Vihear temple, where Thai and Khmer forces stand ready to kill one another if ordered to, soldiers showed me the place where Thai and Khmer forces share holiday celebrations together. *Metta* expresses this relationality between enemies—it is, in fact, a profession of transformative emotions that undermine an individualized self, which became possible because of the conditions of the minefield as a postwar ecology.

The Venerable Maha Ghosananda became the Buddhist Patriarch of Cambodia under the Khmer Rouge regime. Once the genocide ended, despite the continued peripheral battles, he led peace walks on minefields in the countryside. The Khmer Rouge refugee directly connected metta to recall entanglements between enemies. One peace walk route took him and his followers through active battlefields (Skidmore 1996). For this walk, he had a prepared a message:

> Peace is always a point of arrival and a point of departure. That is why we must always begin again, step by step, and never get discouraged. *Bannha* (wisdom) will be our weapons; *metta* (pity-love) and *karuna* (compassion) our bullets; and *satthi* (mindfulness) our armor. We will walk until Cambodia and the whole world is peaceful. (Santidhammo 2009, 25)

Venerable Ghosananda uses the word *metta* as a transformative tool. *Metta* is a bullet, an analogy that implies destroying enemies. Unlike bullets that kill enemies, *metta* destroys enemies in another way, by doing away with them through pity-love and, as my Cambodian informants have suggested to me, by doing away with subjecthood itself through its reciprocity. To understand selves as capable of metta means to let go of a sense of individual selfhood along with righteousness. To let go of the individual, autonomous self, the answer to the question "Man or monster?" evokes shared suffering and accountabilities.

Such an evocation can also be troubling in its implications. We are all men and we are all monsters at once. And despite or because of it, we all deserve metta.

6 Smell (Like) a Rat

Five years after landmine-detection rats arrived in Cambodia, the technique has expanded to over a dozen minefields and platoons in Cambodia. APOPO has even erected a visitor center to provide revenue and publicity for the NGO and to educate tourists on mine-action awareness. The center also serves as a recertification training ground, vacation home, and retirement community for the rats. A simple building, squat and unassuming, it is situated en route to a UNESCO World Heritage Site, a tenth-century temple complex of the Angkor Empire where tourists from all over the world admire moss-covered architecture. The center, however, has more quotidian ambitions for its visitors. Its staff answer questions about rats and their senses.

An APOPO tour guide announces, "We have a contest!" He flourishes his hand and displays a metal tea ball. "Now, you will all know what it is like to be a rat!" Inside the tea ball are fragrant spices, and behind him, a table with a blue cloth on it and ten identical tea balls chained to the wall. On the wall is a photo of a Gambian Pouched Rat in mid-step, her nose bent toward the tea balls.

A pungent spice, earthy and with a sour note, comes through. Its scent is difficult to place. The guide explains, "When you smell this, it will be helpful for you to rub your noses or blow your noses. This is why you see the rats groom their whiskers and rub their faces with their paws. They are clearing the scents off so they can be more precise when they smell." Some of the visitors follow this advice before they approach the table one by one and sniff each tea ball.

Each ball is full of a spice: Kampot peppercorn, Southeast Asian red chili, lemongrass, curry. The visitors smell them multiple times but the intention is to smell each once and then match one of the tea balls to the original. The guide waits until the group is ready to give their answers.

"Okay, so show me in fingers which number ball you think matches the one I gave you."

A kid raises five fingers and most people get this right, though a few raise two hands with different numbers because they're unsure.

"Yes! Five is correct!"

The guide leads a smattering of applause.

"What is the smell? Can anyone tell me?"

A child raises his hand and shouts before he is called, "Cinnamom!"

People laugh at the mispronunciation, but the tour guide nods, "Yes, very good!"

What is it like to be a rat?

The Cambodians I met had answers for what it is like to be a rat based on the postwar ecology that conditioned certain possibilities for their relationships with rats and with each other. Laboring in a platoon of former combatants who now disarm their country of explosive remnants of wars they themselves fought in, their understandings draw from local cosmologies of Theravada Buddhism and local histories of atrocities. Both humans and rats in these relationships must contend with war technologies that configure interspecies and intraspecies relationships. A rat helps two deminers profess love for each other. A ghost kills a rat. Humans tell jokes and secrets while rats find bombs. These stories portray a postwar ecology in the making, a minefield where human and nonhuman beings adapt to new technologies and new relationships. The new choreographies of the landmine detection with rats technique produced models for relationality between former combatants—the ways the deminers loved these rats mediated both the traumatic violence of the past and the uncertain dangers of the minefield.

The deminers and the visitor-center tour guide tell us that they understand the rats via their shared, albeit limited, sensorial encounters of scent detection with the rat. We can approximate what it is like to be a rat because we can distinguish between smells and we can also adore the taste of bananas. Even though humans and rats have varying degrees of such senses and desires, we share them, and, as the guide told me, "This way we can know what it's like to be a rat!" Before, this was found in the shared vulnerabilities of the postwar ecology, but in this contest, shared sensory encounters reveal these odd kinships.

"Since you have succeeded," the guide says after our contest, "we will now ask you to stay and help us in the minefields." He then adds another layer to the ribbing, a seemingly playful gesture toward the politics of mine detection and the realities of child labor: "And we will pay you in bananas."

A murmur of laughter rises at the guide's joke. The contest is not only a game but also an audition. For a moment, the guide asks you to imagine yourself as a rat who can use her sensory powers to find a landmine in the ground—and he tells you that, like the rats, you will be given fruit for finding this military waste. You are confronted with the impossibility of human potential to do so and also marvel that rats are helping humans in such a vital mission. You are also confronted, as I was, with the fact that key to understanding *being* a rat means attending to sensory encounters and how they alter perceptions. This was something that the human deminers practiced when working with the rats, and it became a research practice for me. Being a rat meant being a rat in a postwar ecology.

The visitor center asks us to consider what it is like to be a rat but in a very specific way, framed by landmine-detection techniques. Rather than imagining the typical rat lifestyle—what is it like to forage in meadows at night, to live in underground burrows, or, if we are urban rats, what is it like to run along sewage lines, to leap into trash barrels and feast on garbage?—instead, we are asked to imagine being rats who are companions to humans. Landmine detection, with its materialities and techniques, acted as a framing device to produce trained and domesticated rats who are animal aids, so-called technologies in the minefield. This framing device, as I learned throughout my fieldwork in Cambodia, was essential to imaginaries not only of the rats but also of the humans who worked on the minefield. The technique's materialities affected not only the relationships to the rats but the human deminers' relationships to each other.

Military waste configured both human and nonhuman beings on the minefields of Cambodia. Scholars in feminist technology and science studies and scholars of the ontological turn have proposed that nonhumans reframe how people think about *being human* (Tsing 2012; Haraway 2016; Myers 2017; Kohn 2013; Descola 2013). Haraway (2010, 2013) has discussed both technologies and nonhuman animals as relations with humans that radically disrupt people's imaginaries of humans and nonhumans. For example, the microbiome in our guts, she says (2016), shows us that we are in fact multispecies beings. Natasha Myers (2017) has proposed that plants have the potential to "vegetalize" being human, in that plants can be a way for humans to play out new ways of being in the world.

These calls to reframe what it means to be human are made especially urgent when confronted with the problem of the Anthropocene. They are offered as solutions or alternatives to surviving global climate change through reframing our passive acceptance of capitalist destruction. When Anna Tsing (2015) follows the commodity chain of the matsutake mushroom, flourishing in the ruins of the forest, she offers it as a mode of being that can sidestep capitalist pressures of commodities—ways to live after the ruins of late capitalism destroy human life as we know it. The story of the landmine-detection rats with human deminers offers a narrower scope: the crisis of war legacies. The scope, though, shows how the final logical manifestation of capitalist destruction is militarized ecologies. Because of the presence of landmines, the human and nonhuman beings transformed themselves and their relationships with each other. Because of the humans, rats transformed from pest to companion. Because of the rats, deminers, former combatants, transformed from enemies to colleagues. The counterprogramming of these militarized ecologies can be achieved only through these aspirational postwar relations.

At another stop during the tour is a demonstration of actual landmine detection-rat choreography. A deminer brings the rat out, cradling her like a baby. The deminer wears a government-issued uniform: brown culottes, black boots, and pale blue, button-down shirt with a red Cambodian Mine

Action Center (CMAC) logo patch. The uniform has a military style because government deminers are part of the Cambodian military. Another deminer follows the first with a rake and tea ball in hand. While the one holding the rat waits, the second places the tea ball in a shallow pit of sand. The tourists watch from the sidelines as the tour guide motions for the deminer and the rat to step forward. If any of the tourists reached forward, they would be close enough to touch the rat, but no one does, stayed by the rope that separates the pit from the walkway.

The tour guide introduces the rat, but not the human. "This is Janthina. She is a rat who normally works in the Preah Vihear minefields, at the border of Thailand. She has found fifty-two landmines in three years."

The deminer steps to the outside of the sandpit. Janthina is secured to a metal spring that slides along a line of twine with a pink harness. The deminer steps into a loop on the end of the twine and another deminer stands across the pit, securing the other end in the same way. Janthina walks between the two. For a few minutes, the tour guide is silent. We watch the rat do her work.

Once she approaches the spot where the tea ball has been buried, Janthina furiously rummages in the sand with her forelegs. After her second swipe, one of the deminers clicks a blue clicker, one that is used for animal training, and the rat rushes back to the deminer. The deminer rewards the rat by handing her a half-peeled miniature banana.

The visitors murmur, some saying, "Awww . . ."

The guide explains, "When Janthina smells TNT, the explosive powder in a landmine, she scratches twice."

The rat is not what you expect from a rat, but a rat who has been reconfigured as part of the human deminer team. Each team member is connected to the other by string and the rat runs between the two humans. Our APOPO guide asks us to imagine being a specific kind of rat: a rat employed by the NGO in this team.

"We used to have a rat here called Imani," Meas tells me, grinning next to their girlfriend, Mao. We sit under the banana tree behind the gate that partitions the rat kennels from the visitors' path. Janthina sniffs and wanders in the dirt under the tree on a leash, stretching her legs before the demonstration for the tourists. "Imani was so good for visitors because she used to grasp the tea ball in her hands." Meas demonstrates with their own hands, clutching at the air and scrunching up their nose to mimic a rat. "And she used to bring us the tea ball."

"Yes, but we can't use her," Mao says, shaking her head. "When we held her to show the visitors, she would always bite!"

Imani, then, was better suited to the minefield. Her nibbles would disrupt the play put on for visitors. Even though she fetched the tea ball in the sandpit like a dog, she bit like a rat. The rats' personalities, preferences, and relationships with humans determined the best job for them in Cambodia. They were individuals even though they were also beings whose lives could be imagined through materialities and other sensorial encounters. Their personalities as well as their talents and equipment determined their future roles in the organization. Even though the rats are produced by postwar ecologies, they manage to push against them and negotiate their roles.

The mural on the visitor center's wall proclaims in bold royal blue, yellow, and orange the places where APOPO's activities take place (figure 10). The activities are represented chronologically by the dark silhouettes of rats, exploring painted landmines or bomblets with their noses or sitting up on their hind legs with a graduation cap. According to the mural, APOPO has platoons deployed in Vietnam, Thailand, Cambodia, Colombia, Zimbabwe, and Angola, among others. Each country is accompanied by a year, indicating a weblike timeline of when the platoons have been deployed. The list of countries surprised me, because I knew for a fact that the rats had not been implemented in that many countries. In a misleading ploy that was corrected once I asked, the NGO has sent deminers and equipment to all of these areas, but not rats.[1] The mural uses a rat to represent all the NGO activities, whether or not a rat was part of those activities.

The misleading nature of the mural, however, also proclaims a speculative future for the NGO. The mural has blank spaces where countries can be added, following vein-like yellow stripes. The mural follows other narratives in the NGO's materials and a series of research articles about the landmine-detection-rat technologies. The rats already also detect tuberculosis in sputum samples with a better accuracy rate than microscopic detection, and recent research articles indicate they are even being tested for search and rescue (Poling et al. 2015; Viguers and John 2018; Fiebig

Figure 10. A mural at the APOPO Visitor Center, Siem Reap. Photo by author, 2020.

et al. 2020; Wetzel 2022). There has been a series of news articles about APOPO's partnership with the Endangered Wildlife Trust, a South African–based non-profit, and their 2020 agreement with the Pangolin Crisis Fund. These articles promote the promise of training the rats for finding poached goods like pangolin scales, tobacco, and elephant tusks. The NGO is still training the rats to do this work, sketching visions of the future for rats to become a household name—a companion animal that is as common as a dog, effectively changing the human relationship to rats more generally.

A few rats have made their way to zoos in the United States, like the Tacoma, Washington, Point Defiance Zoo, where three rats demonstrate their training in much the same way the ones at the Siem Reap APOPO visitor center do (Banse 2019). The NGO has transformed the rat from a pest to a companion, rescuer, and even endangered animal protector. The plans for the rats as rescuers in earthquakes and as finders of poached animals are not presently being implemented, but they speak to a future of unlimited potential. These transformations depend on the materialities of landmine detection and the sensorial encounters of these rats and their human handlers combined. At the heart of such materialities is human violence. Wars have made the humans and the rats on a minefield. But the rat is not merely a passive cipher; in fact, the rat has reconfigured these relations in various ways, as have the other beings on the minefield, such as the ghosts and humans.

The story of the landmine-detection rat as she moves from the battlefield to the zoo is a common story of military technology. Technologies developed for war and postwar problems become commodified or reemerge in new settings. This speaks to the dependence that human infrastructures have on military design, such as the selling of leftover US tanks to US police departments. It also speaks to the centers of innovation that military industries have become, from the invention of synthetic fleece and the internet to the development of new biomedical treatments. This is a circumstantial contingency that relies on steady funding from state and corporate leaders. That is, there are more ethical ways to fund human innovations, but generally, humans rely on the military industries to jumpstart novel technologies. For my work, this means that we cannot understand human relations without understanding warfare and that, in the field, these relations push back against military origins in multiple ways.

Postwar technologies push back against their origin stories, just as the people implementing these technologies do not re-create carbon copies of their original prerogatives. They take part in the worlding process through various sensorial encounters.

Take the rat. Take postcolonial, postwar Cambodia. Take a Theravada Buddhist setting and the people who wrangle with working with their former enemies. New materialities and their diverse sensorial encounters allow technologies to exceed their military origin stories. The words Ravi told me come to my mind: "At first, I thought of rats as pests but now I think of them as my very best friends."

When I tried to inquire about what happened to rats when they retired, I encountered contradictory responses. For instance, APOPO hosted an Ask Me Anything (AMA) Reddit thread in the early 2010s. When asked what happened when rats get too old to work (after about eight years), the APOPO manager explained that they live out the rest of their lives in retirement. When I spoke with APOPO members who had worked in Tanzania, where rats were trained at the NGO's headquarters, they explained, "Oh, it's very sad to see the pile of rats when they are retired. They get euthanized." I also knew that the landmine-detection dogs were euthanized at the end of their working lives because it was far too expensive for an NGO to justify the cost of continuing to feed them.[2] But APOPO's act of transforming the rats, which occurred out of necessity to explain the strangeness of the technique and to adapt the misfit rat to the military ethos of the landmine-detection industry, has allowed new possibilities.

One rat, Marcous, of the original platoon, developed skin cancer. Marcous became sluggish at first and unusable for landmine detection. Meas, his handler, felt the hard nodule beneath the fur. "I went to Tok to see," Meas said. Tokla, although a vet, was unsure of the bump's cause, and so Marcous traveled by van to another vet in Phnom Penh. This vet diagnosed Marcous with skin cancer.

In 2016, when Marcous received his diagnosis, the APOPO visitor center was not yet open. The deminers and the supervisors did not want to euthanize him—there were so few of the rats in Cambodia. There was no danger of any rat in Cambodia reproducing because they had all been desexed. Perhaps Marcous could live in comfort?

"So, they contacted me," Ségolène, a long-term resident of Siem Reap, explained to me, "because they knew I would have connections to help the poor little creature." Ségolène was a local celebrity among foreign residents of Siem Reap. She had founded an NGO dedicated to the rescue of stray cats who lived in the streets and temples of the city. A French immigrant to Japan, she retired to Cambodia and lived with a Cambodian family she knew in a nearby village, but she also had an apartment in Siem Reap. She had a bicycle that a local artist had painted with leopard spots, and she proudly called herself a "cat lady," but, as she told me in an interview, she loved all animals. Thin, with a pixie haircut, she often widened her eyes when making a point about the ways in which animals were mistreated in Cambodia.

"So I made sure that the workers at the NGO got in contact with Fede, who runs animal care activities for pets of expats in the city. And he ensured that the sweet little rat could live out his retirement at one of the fanciest hotels and have the fanciest foods."

Every Sunday, Fede organized get-togethers for the foreign residents who had pets. He hired a Cambodian man to walk Marcous on a leash. This was an early antecedent of the visitor center. Locals in the know would come to see the rat as he sniffed the ground and groomed himself. Marcous was popular in the immigrant community—a group of rich foreign residents who lived in Siem Reap and invested in its businesses, worked for NGOs, or benefitted from the high value of their foreign currencies to live a luxurious life they could not afford in their home countries. The popularity of the rat among foreigners pointed to potential for APOPO. Marcous's skin cancer and early retirement promised a new future for the rats. Rat futures were transformed through their new homes and through their effect on their companions via a too-hot sun that killed them before their time. Their human companions, the deminers, did not want to euthanize their new best friend.

The arrival of APOPO inspired new ways of thinking about mine-detection animals that became possible because of the rats. The NGO undermined the militarism among the deminers who worked with former enemies; they reconfigured practices of landmine detection; they connected residents of Siem Reap across class and nationalistic divides; and they traversed oceans as well as times of peace and war.

What is it like to be a rat?

We know what it's like to be the rats in this book by an assemblage of war and postwar technologies. These are particular beings composed of particular concepts and materialities. And we know the rats in this book by their names and the ways they responded to their humans and their training: Victoria, who ran away from Cambodian humans; Simon, who snuggled between two human lovers; Merry, who squeaked while he dreamed; Håvard, who bit a lot of fingers and who was murdered by the spirits; and Marcous, who lived out his early retirement surrounded by adoring fans.

The rats have begun to shape different futures for themselves even in the APOPO visitor center. No longer are the rats limited to HeroRATS, because a new category of rat has emerged: RescueRATS. In photographs these rats wear black and white paneled vests that evoke astronauts, and atop their backs sit video cameras. These rats, like the HeroRATS in Cambodia, will also see for humans, but this time through their capacities to enter into tight spaces in addition to detecting scents, which will guide them to the scent of humans trapped under rubble, to poached animal products, or to drugs.

What then, is it like to be a rat handler?

We know what it's like to be a rat handler from how the humans touched the rats, saw each other, and heard and lived with rumors of violence. We know that fear and love were important emotions on the minefield and structured human and nonhuman relations alike. We know them by their names and their statements and the ways they respond to the rats and to the bombs.

It is interesting to compare ratlike agents with their human companions. The humans shape their futures, too, and they work with the rats and the bombs to imagine it. Landmine clearance is, by the nature of its labor, work toward the future. Deminers seek to transform the land that surrounds them, a land laden with war. The workers and organizations of landmine detection have histories of militarism and violence, but when they worked with these rats, they had opportunities to express love not only for the rats but also for each other. And this love sketched a future of demilitarization and peace in postwar Cambodia. A postwar ecology is an aspirational ecology shaped also by human hope.

When I think of this future, I cannot help but think of Moch and Sovannah's courtship. I remember a photograph where Moch holds her favorite rat, Simon, and Sovannah has his hand on Moch's shoulder, standing close to her. The shot seems candid, and Sovannah and Moch smile at the fat rat in her hands as she cradles him as if he is her baby. Both humans have tilted their heads, lowering their gazes at Simon's nose. The photograph has frozen Moch's hand in mid-caress over Simon's neck. This image stays with me. It has a sense of hopefulness and play that has little to do with the warrior images often portrayed by landmine-detection industries. Instead, its hope comes from the bonds between beings.

War is integral to understanding them, but it is not the only thing. I hope not.

Like the APOPO tour guide, I ask you to imagine what it is like to be on a rat and human deminer team (I leave which being up to you) and, like the tour guide, I ask that this imaginary be circumscribed by technologies of war and postwar. While such temporal distinctions are false oppositions, the bombs of the minefield harken back to old enmities that the deminers must overcome to work together and decontaminate the country of military waste. Postwar is the human aspiration of decontamination. And the rats were integral to these hopes. Rats and string mediated these violent pasts and transformed them: from enemy to coworker, from pest to companion, from feared to beloved.

Notes

CHAPTER 1

1. In this book, I will be using third-person pronouns (*she* or *he*) for the rats rather than object pronouns (*it*) because the humans who engaged with them used the same. In addition, the rats were persons in the minefield and, as I would for any companion animal such as a dog or a cat, I use third person pronouns. By academic conventions I sometimes use she, her, hers for the default pronoun for the rats unless it confuses the subjects and I also use he, him, his when the rat has been identified as male.

2. All names and identities in this book have been changed to protect the humans who consented to tell me about their lives and thoughts for my research. This means that I have also changed the names of the rats. The rats are public figures and often associated with their handlers in the press. Instead of merely using pseudonyms, I have also created composite characters and fictionalizations. Everything written in this book happened as I have described it, but the people represented are not the same people who were part of the stories; instead, they are individuals I knew whose stories and actions I have amalgamated in order to best disguise their identities.

3. Ben Kiernan's book (2002) discusses the range of death toll counts, and Alexander Laban Hinton (2004) also discusses the controversy over such numbers. The numbers of people who died after mass atrocity are contested for a

variety of reasons across the world as Diane Nelson (2015) outlines in her analysis of the death tolls after Guatemala's civil wars.

4. For more details on the conflicts and the governance of Cambodia, see Kiernan 2004; Hinton 2005; and Bultmann 2016.

CHAPTER 2

1. The title for a monk.

2. Literally, "Ask merit enter [monkhood because] what do here." The context of the conversation provided the better translation as well as the past tense, because he was describing his life from 1975 to 1980.

3. This quote is from the testimony of deminer Aki Ra (2017) on his website in support of the NGO he cofounded, Self-Help Demining. It is very similar to the story he told me in person. It was very rare in my fieldwork for people to be this up front and provide such detailed information about the horrors of the past unless they were describing their own suffering. Even in this testimony, he is not describing an act he has undertaken himself. Aki Ra is something of a local celebrity and the NGO depends on his stories and military experience.

4. This is included in tribunal documents and press about the Khmer Rouge Tribunal: https://vodenglish.news/draft-law-passed-to-sunset-the-khmer-rouge-tribunal/. In Khmer the phrase is ខ្មែរសម្លាប់ខ្មែរ.

5. He spoke to me in English and in fact, refused to speak Khmer to me. He sometimes talked about himself in third person, a common thing for Khmer people to do, especially if they are trying to translate their honorifics directly into English. In Khmer, sometimes people of a higher rank will refer to themselves in third person to reinforce the ranking system and also for clarification in speech where pronouns are dropped most of the time.

6. Because he seemed to be showing off, I was not even sure that Brahm's claim to own land was legitimate.

7. In the car, we spoke a combination of Khmer and English because the Tanzanian consultants who also rode to the minefield to train the Khmer staff on the rats did not speak Khmer. Sometimes they fell asleep. When they did, Chann would switch to Khmer with me because he knew I liked to speak Khmer. When the Tanzanians were awake, though, he switched back to English to be polite and include them in the conversations.

CHAPTER 3

1. For a catalogue of landmine-detection personal protective equipment, please see the Geneva International Centre for Humanitarian Demining, https://www.gichd.org/en/resources/equipment-catalogue/.

2. The history of the rat is embedded in a laboratory setting, in both its biological attributes through its breeding and its social connotations. The laboratory rat in psychiatric geneticists' experiments offers a way for scientists to "model" human behaviors and genes (Nelson 2018). Rats become ways to produce knowledge about being human because they can become analogous to humans in terms of genes and behavior. Nelson calls this "epistemic scaffolding."

3. Most landmines in Cambodia were laid in the 1980s during the Vietnamese takeover of the country, which came after the defeat of Pol Pot's Maoist-communist Khmer Rouge during the Cambodian-Vietnamese War (1979–1989). Other munitions are explosive relics of the Vietnam War (which the Vietnamese call "the American War"), when the United States dropped bombs on communist forces. These conflicts are entangled: the US intrusion in Vietnam in the 1960s led to the rise of the Khmer Rouge, a genocidal regime responsible for murdering millions of Cambodians in the 1970s. Although Vietnam and Cambodia were initially united, the two communist regimes soon turned against each other, leaving China to mediate. The result was the K5 belt, an invisible wall preventing Khmer Rouge troops from returning to Cambodia via Thailand. See the introduction and chapter 2 for more details of this history.

4. I have also been told that the Thai soldiers are of Northern Khmer ethnicity and therefore not as separated by nationality as uniforms would suggest.

5. Eleana Kim (2016) develops the concept of "rogue infrastructures" for this paradoxical productive/destructive force of landmines.

6. Scholars and deminers have told me stories of live bombs that remained unexploded even after they traveled long distances. One scholar told me that a professor had taken a bomb from Southeast Asia all the way to his office in the United Kingdom. It had sat on his desk until a general in the army had gone to his office and noticed it was live. Such stories warrant analyses of the parallels I see with the uncertainty of the invisible explosives beneath a serene landscape. The bombs themselves are mysterious and *hold* you. The stories also speak to specialized knowledge—that professional expertise is needed to uncover their level of danger.

7. This did not mean that NGO staff were all scientists. Most of them, even Paul, the New Zealand consultant who boasted that APOPO was "demilitarizing" mine clearance, were veterans with combat experience.

8. *Barang* is also the word for "French" and reflects colonialism in the understanding of skin color. It is also sometimes used interchangeably for "European."

CHAPTER 4

1. I use *ghost* and *spirit* interchangeably in this chapter because, although there are multiple taxonomies in Khmer and other Southeast Asian representations of

the spirits, ranging from monstrous creatures to benevolent demigods, they are all under the umbrella of spiritual entities that do not exist in dichotomous categories. Choosing the word *ghost* for a certain kind of being and *spirit* for another would enforce a dichotomous characterization when, according to the Cambodians I spoke with, these categories are not dichotomous and these entities, even when characterized under a taxonomy of "bad," could be transformed as "good." In fact, the history of the division of categories of good and bad spirits or of spirits versus ghosts is tied up in colonial and Christian understandings of otherworldly beings. Anthropologists Lisa Arensen (2017, 2020) and Krisna Uk (2016) connect these spirits to material practice and landscape understandings as well.

2. *Neak* is a word that means both "you" and "person" although it never means "he," "she," or "it." Likewise, *ta* and *yeay* just mean "grandfather" and "grandmother," so the names for the spirits translate the same as if the speaker were speaking about any living elderly person (all elderly people, regardless of kinship are called "grandparents"). Like the spirits of the forest in the Amazon, the Cambodian and Vietnamese war spirits counter the finality of death and embed the self in a continuity of existence that cannot be defined by one-to-one semiosis. For Kohn (2013), the Amazonian spirits represent signs (in the Peircean sense of the word) of a potential future self. Spirits allow the Runa to "surviv[e] as an I, . . . as it plays out in this ecology of selves . . . [whether] human or nonhuman, fleshy or virtual" (218). Thus, in Kohn's view, anthropologists must understand spirits as real, which "includes but also goes beyond the fact that other people take them to be real, that [they] should take that fact seriously and that we should even be open to how these kinds of real might affect us" (217) because of their importance in the semiotic reality of the world. Kohn's call takes anthropology beyond the human, which, for him, translates to beyond the symbolic—that is, meaning-making defined in reference to other symbols. In their arbitrary nature, symbols cannot totally capture the reality of spirits. Kohn says that all "living selves" exist in the logic of Peircean semiotics because they live *in futuro*— that is, a self is only a potential perspective. To be a self, a person must have someone else that potentially recognizes her as a self, as a *you.*

3. The scholar of Thailand Mue Meut (2012), who uses a pseudonym, depicts modes of understanding hierarchy of both governance and spirits as conspiracy theories that parallel secret shamanic yantra tattoos. He suggests that surveillance, magic, and politics are conceptual assemblages in Southeast Asia.

4. Ashley Thompson's (2020) study of statues who transform from an Angkorian sovereign, Jayavarum VII, to Buddha, as well as the gender fluidity of Buddhist statues, illuminates how such practices shape power structures and underpin concepts of being human. Heonik Kwon (2008) has detailed how spirits can transform in Vietnam, interviewing a medium who, possessed by the spirit of a dead girl, described going to school to become a goddess. In Cambodia, too, Anne Yvonne Guillou (2017), has described ghosts of Khmer Rouge soldiers who

had "bad deaths" from warfare who have begun to move toward a kind of deity status. And Sakada Sakhoeun (2020), a public intellectual and archaeologist in Cambodia, has portrayed an example of a grandmother forest spirit being transformed through prayer/request into a "Grandfather Safety" spirit who can protect people who cross a recently paved road on Phnom Kulen just outside Siem Reap. Such mutabilities not only render dichotomous taxonomies imprecise but also frame ideas of permanence and ordering principles as impositions.

5. Carolyn Culbertson (2013) explores how Judith Butler relates to relationality as a move to reframe ethical anxiety and existentialism.

CHAPTER 5

1. Thanks to a reviewer for this helpful phrase.

2. The crimes of the Anthropocene, variably called the "Capitalocene" or the "Plantationocene" (Davis et al. 2019)—that is, the excesses of human presence on the planet that have led to increased greenhouse gases and climate disasters like fire, tornados, and other eco-horrors (Wald 2022)—are seen as the primacy of the person as individual over all else. A person looks out only for themself.

3. It is possible to imagine that the women who called giant male rats "little sister" used the term of affection not only to assert their kinship with them but also to undermine the fixed connection between body and gender underlying the common assumption that men make better deminers.

4. While presumably a woman could scratch an assailant as well as her lover, the context of "little scratches" made it clear to me that the comment referred to sexual pleasure, not pain.

CHAPTER 6

1. This conflation has confused reporting about APOPO, too, so that some articles claim APOPO's rats had been implemented years earlier than they were in certain countries. The rats, for example, arrived in Cambodia in 2015, but some reports claim they were there in 2014 because the country did have human-only APOPO platoons deployed at that time.

2. There are, however, some singular exceptions, such as the director general's wife who adopted a dog about to be retired.

References

Anjum, Amna, Xu Ming, Ahmed Faisal Siddiqi, and Samma Faiz Rasool. 2018. "An Empirical Study Analyzing Job Productivity in Toxic Workplace Environments." *International Journal of Environmental Research and Public Health* 15, no. 5: 1035.

Antelme, Michel. 2007. *Inventaire provisoire des caractères et divers signes des écritures khmères pré-modernes et modernes employés pour la notation du khmer, du siamois, des dialectes thaïs méridionaux, du sanskrit et du pali.* Projet corpus des inscriptions khmères. http://www. aefek. fr/wa_files /antelme_bis.pdf.

Arensen, Lisa J. 2017. "The Dead in the Land: Encounters with Bodies, Bones, and Ghosts in Northwestern Cambodia." *The Journal of Asian Studies* 76, no. 1: 69–86.

———. 2020. "Shifting Technologies, Mines, and Rituals of Relatedness: A Multimedia Montage of Post-war Cambodia and Laos." Panel presentation at Association for Asian Studies in Asia, online.

Århem, Kaj. 2015. "Southeast Asian Animism in Context." In *Animism in Southeast Asia,* edited by Kaj Århem and Guido Sprenger, 3–29. Routledge.

Asad, Talal. 2007. *On Suicide Bombing.* Columbia University Press.

An, Sokhieng. 2012. *Mixed Medicines: Health and Culture in French Colonial Cambodia.* University of Chicago Press.

Aulino, Felicity. 2012. "Senses and Sensibilities: The Practice of Care in Everyday Life in Northern Thailand." PhD diss., Harvard University.

———. 2019. *Rituals of Care: Karmic Politics in an Aging Thailand*. Cornell University Press.

Bach, Håvard, and Ian McLean. 2003. "Remote Explosive Scent Training: Genuine or a Paper Tiger?" *Journal of Conventional Weapons Destruction* 7, no. 1: 75-82.

Banse, Tom. 2019. "'Hero Rats' Arrive At Tacoma Zoo. Their Relatives Are Still Saving Lives." In the News, OPB, July 20, 2019. https://www.opb.org/news/article/tacoma-zoo-hero-rats-landmine-detection-training.

Barthes, Roland. 1977. *Image-Music-Text*. Macmillan.

Bateson, Gregory. 2000. *Steps to an Ecology of Mind: Collected Essays in Anthropology, Psychiatry, Evolution, and Epistemology*. University of Chicago Press.

Beban, A., and L. Schoenberger. 2019. "Fieldwork Undone: Knowing Cambodia's Land Grab through Affective Encounters." *ACME: An International Journal for Critical Geographies* 18, no. 1: 77–103.

Beban, Alice, and Courtney Work. 2014. "The Spirits Are Crying: Dispossessing Land and Possessing Bodies in Rural Cambodia." *Antipode* 46, no. 3: 593–610.

Benson, Etienne. 2013. "The Urbanization of the Eastern Gray Squirrel in the United States." *Journal of American History* 100, no. 3: 691–710.

Berlant, Lauren. 2012. *Desire/Love*. Punctum Books.

Bessire, Lucas. 2017. "On Negative Becoming." In *Unfinished: The Anthropology of Becoming*, edited by João Biehl and Peter Locke. Duke University Press.

Bettache, Karim. 2020. "A Call to Action: The Need for a Cultural Psychological Approach to Discrimination on the Basis of Skin Color in Asia." *Perspectives on Psychological Science* 15, no. 4: 1131–39.

Bisang, Walter. 2014. "On the Strength of Morphological Paradigms." In *Paradigm Change: In the Transeurasian Languages and Beyond*, edited by Martine Robbeets and Walter Bisang, 23–60. John Benjamins Publishing Company.

Bjork-James, Sophie. 2018. "Training the Porous Body: Evangelicals and the Ex-Gay Movement." *American Anthropologist* 120, no. 4: 647–58.

Bottomley, Ruth. 2003. *Crossing the Divide: Landmines, Villagers and Organizations*. International Peace Research Institute.

Brickell, Katherine. 2020. *Home SOS: Gender, Violence, and Survival in Crisis Ordinary Cambodia*. John Wiley & Sons.

Broglio, Ron. 2013. "When Animals and Technology Are Beyond Human Grasping." *Journal of the Theoretical Humanities* 18, no. 1: 1–9.

Brown, Wendy. 2005. *Edgework: Critical Essays on Knowledge and Politics*. Princeton University Press.

Bultmann, Daniel. 2012. "Irrigating a Socialist Utopia: Disciplinary Space and Population Control under the Khmer Rouge, 1975-1979." *Transcience* 3, no. 1: 40–52.

———. 2015. *Inside Cambodian Insurgency: A Sociological Perspective on Civil Wars and Conflict*. Ashgate Publishing.

Butler, Judith. 1995. "Collected and Fractured: Responses to Identities." In *Identities* edited by Anthony Appiah and Henry Louis Gates. University of Chicago Press.

Candea, Matei. 2010. "'I Fell in Love with Carlos the Meerkat': Engagement and Detachment in Human-Animal Relations." *American Ethnologist* 37, no 2: 241–58.

Cassaniti, Julia. 2015. *Living Buddhism: Mind, Self, and Emotion in a Thai Community*. Cornell University Press.

Cavell, Stanley. 1988. "Declining Decline: Wittgenstein as a Philosopher of Culture." *Inquiry* 31, no. 3: 253–64.

Chandler, David. 1982. "The Assassination of Resident Bardez (1925): A Premonition of Revolt in Colonial Cambodia." *Journal of the Siam Society* 70, no. 1: 35–49.

Chhim, Sotheara. 2012. "Baksbat (Broken Courage): The Development and Validation of the Inventory to Measure Baksbat, a Cambodian Trauma-Based Cultural Syndrome of Distress." *Culture, Medicine, and Psychiatry* 36: 640–59.

Choulean, Ang. 1988. "The Place of Animism within Popular Buddhism in Cambodia: The Example of the Monastery." *Asian Folklore Studies* 47, no. 1: 35–41.

———. 2005. *Brah Ling*. Reyum Institute of Arts and Culture.

Coetzee, John Maxwell. 2016. *The Lives of Animals*. Vol. 43, University Center for Human Values Series. Princeton University Press.

Coleman, Gabriella. 2014. *Hacker, Hoaxer, Whistleblower, Spy: The Many Faces of Anonymous*. Verso Books.

Croissant, Aurel, and Philip Lorenz. 2018. "Cambodia: From UN-Led Peace-Building to Post-Genocidal Authoritarianism." In *Comparative Politics of Southeast Asia*, Springer Texts in Political Science and International Relations, 35–69. Springer. https://doi.org/10.1007/978-3-031-05114-2_3.

Culbertson, Carolyn. 2013. "The Ethics of Relationality: Judith Butler and Social Critique." *Continental Philosophy Review* 46, no. 3: 449–63.

Cummings, Joe. 2012. *Sacred Tattoos of Thailand: Exploring the Magic, Masters, and Mystery of Sak Yan*. Marshall Cavendish International (Asia) Pte.

Curley, Melissa. 2018. "Governing Civil Society in Cambodia: Implications of the NGO Law for the 'Rule of Law.'" *Asian Studies Review* 42, no. 2: 247–67.

Crane, Emma Shaw. 2023. "Lush Aftermath: Race, Labor, and Landscape in the Suburb." *Environment and Planning D: Society and Space* 41, no. 2: 210–30.

Dave, Naisargi N. 2022. "Love and Other Injustices: On Humans, Animals, and an Ethics of Indifference." *Comparative Studies of South Asia, Africa, and the Middle East* 43, no 3: 656–67.

Davis, Erik W. 2015. *Deathpower: Buddhism's Ritual Imagination in Cambodia*. Columbia University Press.

Davis, Janae, Alex A. Moulton, Levi Van Sant, and Brian Williams. 2019. "Anthropocene, Capitalocene, . . . Plantationocene?: A Manifesto for Ecological Justice in an Age of Global Crises." *Geography Compass* 13, no. 5: e12438.

Deacon, Terrence W. 2011. *Incomplete Nature: How Mind Emerged from Matter*. W. W. Norton & Company.

DeAngelo, Darcie. 2018. "Demilitarizing Disarmament with Mine Detection Rats." *Culture and Organization* 24, no. 4: 285–302.

———. 2019. "Negative Space: Imaginaries of Violence in Cambodia." *Southeast of Now: Directions in Contemporary and Modern Art in Asia* 3, no. 2: 45–64.

DeAngelo, Darcie, and Deborah A. Jones. 2019. "Explosive Landscapes." *Anthropology News* 60, no. 6: e74–e80.

Descola, Philippe. 2013. *Beyond Nature and Culture*. University of Chicago Press.

Despret, Vinciane. 2004. "The Body We Care For: Figures of Anthropo-Zoo-Genesis." *Body & Society* 10, no. 2–3: 111–34.

Drouyer, Isabel Azevedo, Rene Drouyer, Narisa Chakrabongse, and Mōm Rātchawong. 2013. *Thai Magic Tattoos: The Art and Influence of Sak Yant*. Bangkok River Books.

Eberhardt, Nancy. 2006. *Imagining the Course of Life: Self Transformation in a Shan Buddhist Community*. University of Hawai'i Press.

Ebihara, May Mayko. (1968) 2018. *Svay: A Khmer Village in Cambodia*. Cornell University Press.

Edwards, Penny. 2007. *Cambodge: The Cultivation of a Nation, 1860–1945*. University of Hawaii Press.

Eisenbruch, Maurice. 1992. "The Ritual Space of Patients and Traditional Healers in Cambodia." *Bulletin de l'École française d'Extrême-Orient* 79, no 2: 283–316.

Estren, Mark J. 2012. "The Neoteny Barrier: Seeking Respect for the Non-Cute." *Journal of Animal Ethics* 2, no. 1: 6–11.

Fanon, Frantz. 1963. "Colonial War and Mental Disorders." In *The Wretched of the Earth* by Frantz Fanon, translated by Richard Philcox. Grove Press.

Fernando, Mayanthi. 2017. "Supernatureculture." *The Immanent Frame: Secularism, Religion, and the Public Sphere*, December 11, 2017. https://tif.ssrc.org/2017/12/11/supernatureculture/.

———. 2022. "Uncanny Ecologies: More-than-Natural, More-than-Human, More-than-Secular." *Comparative Studies of South Asia, Africa and the Middle East* 42, no. 3: 568–83.

Fiebig, Lena, Negussie Beyene, Robert Burny, Cynthia D. Fast, Christophe Cox, and Georgies F. Mgode. 2020. "From Pests to Tests: Training Rats to Diagnose Tuberculosis." *European Respiratory Journal* 55, no. 3: 1-5.

Forest, Alain. 1993. *Le Culte des Génies Protecteurs au Cambodge: Analyse et Traduction d'un Corpus de Textes sur les Neak Ta.* L'Harmattan.

French, Lindsay. 1994. "The Political Economy of Injury and Compassion: Amputees on the Thai-Cambodia Border." In *Embodiment and Experience: The Existential Ground of Culture and Self,* edited by Thomas J. Csordas Alan Harwood, 69-99. Cambridge University Press.

Fuentes, Agustín. 2010. "Naturalcultural Encounters in Bali: Monkeys, Temples, Tourists, and Ethnoprimatology." *Cultural Anthropology* 25, no. 4: 600-624.

Gell, Alfred. 1998. *Art and Agency: An Anthropological Theory.* Clarendon Press.

Gilley, Bruce. 2020. "Assaults on Capitalism and Democratic Backsliding: Evidence from Asia." *Asian Journal of Comparative Politics* 6, no 3: 293-309.

Ginsburg, Faye, and Rayna Rapp. 2013. "Disability Worlds." *Annual Review of Anthropology* 42, no. 2013: 53-68.

———. 2018. "Cognitive Disability: Towards an Ethics of Possibility." *The Cambridge Journal of Anthropology* 36, no. 1: 113-19.

———. 2020. "Disability/Anthropology: Rethinking the Parameters of the Human: An Introduction to Supplement 21." *Current Anthropology* 61, no. S21: S4-S15.

Govindrajan, Radhika. 2018. *Animal Intimacies: Interspecies Relatedness in India's Central Himalayas.* University of Chicago Press.

Grauerholz, Liz. 2007. "Cute Enough to Eat: The Transformation of Animals into Meat for Human Consumption in Commercialized Images." *Humanity & Society* 31, no. 4: 334-54.

Guillou, Anne Yvonne. 2012. "An Alternative Memory of the Khmer Rouge Genocide: The Dead of the Mass Graves and the Land Guardian Spirits [Neak Ta]." *South East Asia Research* 20, no. 2: 207-26.

———. 2017. "Khmer Potent Places: Pāramī and the Localisation of Buddhism and Monarchy in Cambodia." *The Asia Pacific Journal of Anthropology* 18, no. 5: 421-43.

Gusterson, Hugh. 1996. *Nuclear Rites: A Weapons Laboratory at the End of the Cold War.* University of California Press.

Habib, Maki K. 2007. "Controlled Biological and Biomimetic Systems for Landmine Detection." *Biosensors and Bioelectronics* 23, no. 1: 1-18.

Haiman, John. 1999. "Auxiliation in Khmer the Case of Baan." *Studies in Language* 23, no. 1: 149-72.

Hamilton, Lindsay, and Darren McCabe. 2016. "'It's Just a Job': Understanding Emotion Work, De-Animalization and the Compartmentalization of Organized Animal Slaughter." *Organization* 23, no. 3: 330-50.

Hammond, Susan, and Arnold Schecter. 2012. "Agent Orange: Health and Environmental Issues in Vietnam, Cambodia, and Laos." In *Dioxins and*

Health: Including Other Persistent Organic Pollutants and Endocrine Disruptors, edited by Arnold Schechter, 469–520. Wiley.

Han, Clara, and Veena Das. 2015. "Introduction: A Concept Note." In *Living and Dying in the Contemporary World,* edited by Clara Han and Veena Das. 1–38. University of California Press.

Hancock, Virginia. 2008. "No-Self at Trial: How to Reconcile Punishing the Khmer Rouge for Crimes against Humanity with Cambodian Buddhist Principles." *Wisconsin International Law Journal* 26: 87-130.

Haraway, Donna Jeanne. 1994. "A Game of Cat's Cradle: Science Studies, Feminist Theory, Cultural Studies." *Configurations* 2, no. 1: 59–71.

———. 2007. *When Species Meet.* University of Minnesota Press.

———. 2010. "A Cyborg Manifesto" (1985). In *Cultural Theory: An Anthology,* edited by Imre Szeman and Timothy Kaposy. Wiley Blackwell.

———. 2013. *When Species Meet,* 3rd ed. University of Minnesota Press.

———. 2016. *Staying with the Trouble: Making Kin in the Chthulucene.* Duke University Press.

Henig, David. 2019. "Living on the Frontline: Indeterminacy, Value, and Military Waste in Postwar Bosnia-Herzegovina." *Anthropological Quarterly* 92, no. 1: 85–110.

Hinton, Alexander Laban. 1998. "A Head for an Eye: Revenge in the Cambodian Genocide." *American Ethnologist* 25, no. 3: 352–77.

———. 2005. *Why Did They Kill?: Cambodia in the Shadow of Genocide.* University of California Press.

———. 2016. *Man or Monster?: The Trial of a Khmer Rouge Torturer.* Duke University Press.

Ingold, Tim. 2012. "Toward an Ecology of Materials." *Annual Review of Anthropology* 41, no. 2012: 427–42.

Jacobsen, Trudy. 2008. *Lost Goddesses: The Denial of Female Power in Cambodian History.* NIAS Press.

Jayawardene, Sureshi M. 2016. "Racialized Casteism: Exposing the Relationship between Race, Caste, and Colorism through the Experiences of Africana People in India and Sri Lanka." *Journal of African American Studies* 20: 323–45.

Khare, R. S. 1995. "The Body, Sensoria, and Self of the Powerless: Remembering/'Re-Membering' Indian Untouchable Women." *New Literary History* 26, no. 1: 147–68.

Khayyat, Munira. 2022. *A Landscape of War: Ecologies of Resistance and Survival in South Lebanon.* University of California Press.

Kiernan, Ben. 2004. *How Pol Pot Came to Power: Colonialism, Nationalism, and Communism in Cambodia, 1930-1975.* Yale University Press.

Kim, Eleana J. 2014. "The Flight of Cranes: Militarized Nature at the North Korea–South Korea Border." *RCC Perspectives* 3, no. 3 (201): 65–70.

———. 2016. "Toward an Anthropology of Landmines: Rogue Infrastructure and Military Waste in the Korean DMZ." *Cultural Anthropology* 31, no. 2: 162–87.

———. 2022. *Making Peace with Nature: Ecological Encounters along the Korean DMZ.* Duke University Press.

Kirk, Robert G. W. 2014. "In Dogs We Trust? Intersubjectivity, Response-Able Relations, and the Making of Mine Detector Dogs." *Journal of the History of the Behavioral Sciences* 50, no. 1: 1–36.

Kitagawa, Takako. 2005. "Kampot of the Belle Époque: From the Outlet of Cambodia to a Colonial Resort." *Japanese Journal of Southeast Asian Studies* 42, no. 4: 394–417.

Kleinman, Arthur. 1995. *Writing at the Margin: Discourse between Anthropology and Medicine.* University of California Press.

Klima, Alan. 2009. *The Funeral Casino.* Princeton University Press.

Kockelman, Paul. 2010. "Enemies, Parasites, and Noise: How to Take Up Residence in a System without Becoming a Term in It." *Journal of Linguistic Anthropology* 20, no. 2: 406–21.

Kohn, Eduardo. 2013. *How Forests Think: Toward an Anthropology beyond the Human.* University of California Press.

Kuriyama, Shigehisa. 1999. *The Expressiveness of the Body and the Divergence of Greek and Chinese Medicine.* Zone Books.

Kwon, Heonik. 2008. *Ghosts of War in Vietnam.* Vol. 27, Studies in the Social and Cultural History of Modern Warfare. Cambridge University Press.

Landmine and Cluster Munitions Monitor. 2016. "Cambodia: Mine Action." International Campaign to Ban Landmines and Cluster Munitions Commission. *Landmine Monitor,* last updated November 22, 2016. http://the-monitor.org/en-gb/reports/2016/cambodia/mine-action.aspx

Latour, Bruno. 2005. *Reassembling the Social: An Introduction to Actor-Network-Theory.* Oxford University Press.

Lê, Viet. 2021. *Return Engagements: Contemporary Art's Traumas of Modernity and History in Sài Gòn and Phnom Penh.* Duke University Press.

Ledgerwood, Judy. 1994. "Gender Symbolism and Culture Change: Viewing the Virtuous Woman." In *Cambodian Culture since 1975 Homeland and Exile,* edited by Carol Anne Mortland, Judy Ledgerwood, and May Ebihara, 119–28. Cornell University Press.

———. 2008. "Buddhist Practice in Rural Kandal Province, 1960 and 2003." In *People of Virtue: Reconfiguring Religion, Power, and Moral Order in Cambodia Today,* edited by Alexandra Kent and David Porter Chandler, 147–168. NIAS Press.

Lee, Jia Hui. 2021. "Rat Tech: Transforming Rodents into Technology in Tanzania." *Arcadia,* no. 5 Rachel Carson Center for Environment and Society. doi:10.5282/rcc/9214.

LeVine, Peg. 2010. *Love and Dread in Cambodia: Weddings, Births, and Ritual Harm under the Khmer Rouge.* NUS Press.

Lim, Samson. 2016. *Siam's New Detectives: Visualizing Crime and Conspiracy in Modern Thailand.* University of Hawaii Press.

Luhmann, Niklas. 1986. *Love as Passion: The Codification of Intimacy.* Harvard University Press.

Luhrmann, Tanya. 2011. "Toward an Anthropological Theory of Mind." *Suomen Antropologi: Journal of the Finnish Anthropological Society* 36, no. 4: 5–69.

Lynch, Michael E. 1988. "Sacrifice and the Transformation of the Animal Body into a Scientific Object: Laboratory Culture and Ritual Practice in the Neurosciences." *Social Studies of Science* 18, no. 2: 265–89.

Lyons, Kristina M. 2020. *Vital Decomposition: Soil Practitioners and Life Politics.* Duke University Press.

MacDonald, Jacqueline, J. R. Lockwood, John McFee, Thomas Altshuler, and Thomas Broach. 2003. *Alternatives for Landmine Detection.* RAND/MR-1608-OSTP. RAND Corporation.

Mahoney, Amanda, Kate Lalonde, Timothy Edwards, Christophe Cox, Bart Weetjens, and Alan Poling. 2014. "Landmine-Detection Rats: An Evaluation of Reinforcement Procedures under Simulated Operational Conditions." *Journal of the Experimental Analysis of Behavior* 101, no. 3: 450–56.

Mbembe, Achille. 2019. *Necropolitics.* Duke University Press.

McLean, Stuart. 2008. "Bodies from the Bog: Metamorphosis, Non-Human Agency and the Making of 'Collective' Memory." *Trames* 12, no. 3: 299–308.

Meut, Mue. 2012. "History of Conspiracy." *Sensate: A Journal for Experiments in Critical Media Practice, Unspeakable Things,* August. http://sensatejournal.com/2012/08/mue-meut-history-of-conspiracy/.

Minh-ha, Trinh (Thi Minh-Ha). 1989. *Woman, Native, Other: Writing Postcoloniality and Feminism.* Indiana University Press.

———. 1992. *Framer Framed.* Psychology Press.

———. 2012. *The Digital Film Event.* Routledge.

———. 2016. *Lovecidal: Walking with the Disappeared.* Fordham University Press.

Mol, Annemarie. 2003. *The Body Multiple.* Duke University Press.

Moten, Fred. 2003. *In the Break: The Aesthetics of the Black Radical Tradition.* University of Minnesota Press.

Munt, Sally. 2003. "The Butch Body." In *Contested Bodies,* edited by John Hassard, and Ruth Holliday, 95–106. Taylor and Francis.

Myers, Natasha. 2017. "From the Anthropocene to the Planthroposcene: Designing Gardens for Plant/People Involution." *History and Anthropology* 28, no. 3: 297–301.

Nading, Alex M. 2014. *Mosquito Trails: Ecology, Health, and the Politics of Entanglement.* University of California Press.

Nagel, Thomas. 1974. "What Is It Like to Be a Bat?" *The Philosophical Review* 83, no. 4: 435–50.

Nakamura, Karen. 2013. "Making Sense of Sensory Ethnography: The Sensual and the Multisensory." *American Anthropologist* 115, no. 1: 132–35.

Nam, Sylvia. 2011. "Phnom Penh: From the Politics of Ruin to the Possibilities of Return." *Traditional Dwellings and Settlements Review* 23, no 1: 55–68.

National Institute of Statistics, Directorate General for Health in Cambodia, and ICF International, 2015. "Cambodia Demographic and Health Survey 2014." Phnom Penh, Cambodia, and Rockville, MD: National Institute of Statistics, Directorate General for Health in Cambodia, and ICF International.

Neef, Andreas, and Siphat Touch. 2012. "Land Grabbing in Cambodia: Narratives, Mechanisms, Resistance." Draft paper for Land Deal Politics Initiative presented at Global Land Grabbing II, Oct. 17–19, 2012. Conference papers published by Cornell University.

Nelson, Maggie. 2016. *The Argonauts*. Melville House.

Nelson, Nicole C. 2018. *Model Behavior: Animal Experiments, Complexity, and the Genetics of Psychiatric Disorders*. University of Chicago Press.

Nhem, Boraden. 2014. *The Third Indochina Conflict: Cambodia's Total War*. Master's thesis, Army Command and General Staff College.

Nittono, Hiroshi, Michiko Fukushima, Akihiro Yano, and Hiroki Moriya. 2012. "The Power of Kawaii: Viewing Cute Images Promotes a Careful Behavior and Narrows Attentional Focus." *PLOS One* 7, no. 9: e46362.

Noordegraaf, Herman. 2018. "Mercy and Justice." In *Considering Compassion: Global Ethics, Human Dignity, and the Compassionate God*, edited by Frits de Lange and L. Juliana Claassens, 145–56. Pickwick Publications.

Nordstrom, Carolyn. 2004. *Shadows of War: Violence, Power, and International Profiteering in the Twenty-First Century*. University of California Press.

Nyanatiloka, Thera. 1997. *Buddhist Dictionary*. Karunaratne & Sons.

Obeyesekere, Gananath. 2002. *Imagining Karma: Ethical Transformation in Amerindian, Buddhist, and Greek Rebirth*. Vol. 14, Comparative Studies in Religion and Society. University of California Press.

Oesterheld, Christian. 2014. "Cambodian-Thai Relations during the Khmer Rouge Regime: Evidence from the East German Diplomatic Archives." *Silpakorn University Journal of Social Sciences, Humanities, and Art* 14, no. 2: 107–28.

Oeur, V. Sokuntevy. 2018. Portfolio. *Visual Anthropology* 31, no. 1–2: 166–69. doi: 10.1080/08949468.2018.1429779.

Pardo Pedraza, Diana. 2023. "Ethical Disconcertment and the Politics of Troublemaking: Land Mines, Humanitarian Demining, and Ecologies of Trouble in Rural Colombia." *American Ethnologist* 50, no 3: 463–73.

Pardo Pedraza, Diana, and Julia Morales Fontanilla. 2023. "Explosiveness: Territories of War and Technoscientific Practices in Colombia." *Journal of Latin American and Caribbean Anthropology* 28, no 3: 239–50.

Parreñas, Juno Salazar. 2018. *Decolonizing Extinction: The Work of Care in Orangutan Rehabilitation.* Duke University Press.

Peou, Sorpong. 2002. "Realism and Constructivism in Southeast Asian Security Studies Today: A Review Essay." *The Pacific Review* 15, no. 1: 119–38.

Pink, Sarah. 2015. *Doing Sensory Ethnography.* Sage.

Pinto-García, Lina. 2022. "Military Dogs and Their Soldier Companions: The More-Than-Human Biopolitics of Leishmaniasis in Conflict-Torn Colombia." *Medical Anthropology Quarterly* 23, no 2: 237–55.

Poling, Alan, Amanda Mahoney, Negussie Beyene, Georgies Mgode, Bart Weetjens, Christophe Cox, and Amy Durgin. 2015. "Using Giant African Pouched Rats to Detect Human Tuberculosis: A Review." *Pan-African Medical Journal* 21, no. 1: 2–6.

Povinelli, Elizabeth A. 2006. *The Empire of Love: Toward a Theory of Intimacy, Genealogy, and Carnality.* Duke University Press.

Prokosch, Eric. 1995. *The Technology of Killing: A Military and Political History of Antipersonnel Weapons.* Zed Books.

Rechtman, Richard. 2006. "The Survivor's Paradox: Psychological Consequences of the Khmer Rouge Rhetoric of Extermination." *Anthropology & Medicine* 13, no. 1: 1–11.

———. 2021. *Living in Death: Genocide and Its Functionaries.* Fordham University Press.

Reid, Anthony. 1993. *Southeast Asia in the Age of Commerce 1450–1680.* Vol. 2, Expansion and Crisis. Yale University Press.

Reno, Joshua. 2016. *Waste Away: Working and Living with a North American Landfill.* University of California Press.

———. 2020. *Military Waste: The Unexpected Consequences of Permanent War Readiness.* University of California Press.

Roberts, Charlotte. 2017. "Applying the 'Index of Care' to a Person Who Experienced Leprosy in Late Medieval Chichester, England." In *New Developments in the Bioarchaeology of Care*, edited by Lorna Tilley and Alecia A. Schrenk, 101–24. Springer.

Rose, Deborah Bird, Thom Van Dooren, Matthew Chrulew, Stuart Cooke, Matthew Kearnes, and Emily O'Gorman. 2012. "Thinking through the Environment, Unsettling the Humanities." *Environmental Humanities* 1, no. 1: 1–5.

Ruddick, Sue. 2015. "Situating the Anthropocene: Planetary Urbanization and the Anthropological Machine." *Urban Geography* 36, no. 8: 1113–30.

Ruiz-Serna, Daniel. 2023. *When Forests Run Amok: War and Its Afterlives in Indigenous and Afro-Colombian Territories.* Duke University Press.

Sakhoeun, Sakada. 2020. "One Road and Three Towers: A Photographic Essay of Technology, Power, and Access on Kulen Mountain." Panel presentation for Association of Asian Studies in Asia, online.

Santidhammo, Bhikku 2009. *Somdech Preah Maha Ghosananda: The Buddha of the Battlefields*. S. R. Publishing. ghosananda.org.

Scheper-Hughes, Nancy, and Margaret M. Lock. 1987. "The Mindful Body: A Prolegomenon to Future Work in Medical Anthropology." *Medical Anthropology Quarterly* 1, no. 1: 6–41.

Sear, Cynthia. 2020. "Porous Bodies: Corporeal Intimacies, Disgust and Violence in a COVID-19 World." *Anthropology in Action* 27, no. 2: 73–77.

Sherman, Gary D., Jonathan Haidt, and James A. Coan. 2009. "Viewing Cute Images Increases Behavioral Carefulness." *Emotion* 9, no. 2: 282.

Simpson, Andrew. 2005. "Pro Drop Patterns and Analyticity." Lecture notes for "Syntactic Analyticity," for the course Dialogues in Grammatical Theory, Experiment, and Change. Linguistic Society of America Institute, July 27–August 5, 2005. Published as Notes from the Linguistic Society of America. Massachusetts Institute of Technology and Harvard University, administered by the Massachusetts Institute of Technology.

Singh, Sarinda. 2012. *Natural Potency and Political Power: Forests and State Authority in Contemporary Laos*. University of Hawai'i Press.

Sivhuoch, Ou, and Kim Sedara. 2013. *20 Years' Strengthening of Cambodian Civil Society: Time for Reflection*. CDRI Working Paper Series No. 85.

Skidmore, Monique. 1996. "In the Shade of the Bodhi Tree: Dhammayietra and the Re-Awakening of Community in Cambodia." *Crossroads: An Interdisciplinary Journal of Southeast Asian Studies* 10, no. 1: 1–32.

Slocomb, Margaret. 2001. "The K5 Gamble: National Defence and Nation Building under the People's Republic of Kampuchea." *Journal of Southeast Asian Studies* 32, no 2: 195–210.

Smith, Shane, Melissa Devine, Joseph Taddeo, and Vivian Charles McAlister. 2017. "Injury Profile Suffered by Targets of Antipersonnel Improvised Explosive Devices: Prospective Cohort Study." *BMJ Open* 7, no. 7: e014697.

Stevenson, Lisa. 2014. *Life Beside Itself: Imagining Care in the Canadian Arctic*. University of California Press.

Stewart, Kathleen. 2007. *Ordinary Affects*. Duke University Press.

Stoller, Paul. 2010. *Sensuous Scholarship*. University of Pennsylvania Press.

Stone, Nomi. 2022. *Pinelandia: An Anthropology and Field Poetics of War and Empire*. Vol. 8, Atelier: Ethnographic Inquiry in the Twenty-First Century. University of California Press.

Strathern, Marilyn. 1987. *Dealing with Inequality: Analysing Gender Relations in Melanesia and Beyond*. CUP Archive.

———. 2005. *Kinship, Law and the Unexpected: Relatives Are Always a Surprise*. Cambridge University Press.

Taksdal, Merete. 2011. "'My Story Started from Food Shortage and Hunger':
Living with Landmines in Cambodia." In *Disability and Poverty: A Global
Challenge,* edited by Arne E. Heide and Benedicte Ingstad, 189–206. Policy
Press.

TallBear, Kim. 2011. " Why Interspecies Thinking Needs Indigenous Stand-
points." *Cultural Anthropology,* Fieldsights, November 18, 2011. The Human
Is More Than Human. https://culanth.org/fieldsights/why-interspecies-
thinking-needs-indigenous-standpoints.

———. 2017. "Beyond the Life/Not-Life Binary: A Feminist-Indigenous Reading
of Cryopreservation, Interspecies Thinking, and the New Materialisms." In
Cryopolitics: Frozen Life in a Melting World, edited by Joanna and Emma
Kowal Radin. MIT Press.

Tambiah, Stanley Jeyaraja. 1984. *The Buddhist Saints of the Forest and the Cult
of Amulets: A Study in Charisma, Hagiography, Sectarianism, and Millen-
nial Buddhism.* Cambridge University Press.

Taussig, Michael T. 1999. *Defacement: Public Secrecy and the Labor of the
Negative.* Stanford University Press.

Taylor, Charles. 2009. *A Secular Age.* Harvard University Press.

Theidon, Kimberly. 2013. *Intimate Enemies: Violence and Reconciliation in
Peru.* University of Pennsylvania Press.

Thompson, Ashley. 2004. "The Suffering of Kings: Substitute Bodies, Healing,
and Justice in Cambodia." In *History, Buddhism, and New Religious
Movements in Cambodia,* edited by John Marston and Elizabeth Guthrie,
91–112. University of Hawaii Press.

———. 2013. "Forgetting to Remember, Again: On Curatorial Practice
and 'Cambodian Art' in the Wake of Genocide." *Diacritics* 41, no. 2:
82–109.

———. 2016. *Engendering the Buddhist State: Territory, Sovereignty and
Sexual Difference in the Inventions of Angkor.* Routledge.

———. 2020. "Anybody: Diasporic Subjectivities and the Figure of the 'Historical' Buddha." In *Interlaced Journeys: Diaspora and Contemporary Southeast Asian Art,* edited by Patrick D. Flores and Loredana Pazzini-Paracciani,
113–28. Osage Publications.

Throop, C. Jason. 2010. "Latitudes of Loss: On the Vicissitudes of Empathy."
American Ethnologist 37, no. 4: 771–82.

Ticktin, Miriam. 2016. "What's Wrong with Innocence." *Cultural Anthropology,*
Fieldsights, July 28, 2016. Refugees and the Crisis of Europe. https://
culanth.org/fieldsights/whats-wrong-with-innocence.

———. 2017. "A World without Innocence." *American Ethnologist* 44, no. 4:
577–90.

Tsing, Anna Lowenhaupt. 2005. *Friction: An Ethnography of Global Connection.* Princeton University Press.

———. 2012. "Unruly Edges: Mushrooms as Companion Species for Donna Haraway." *Environmental Humanities* 1, no. 1: 141–54.

———. 2015. *The Mushroom at the End of the World: On the Possibility of Life in Capitalist Ruins*. Princeton University Press.

———. 2016. "Earth Stalked by Man." *The Cambridge Journal of Anthropology* 34, no. 1: 2–16.

Uk, Krisna. 2016. *Salvage: Cultural Resilience among the Jorai of Northeast Cambodia*. Cornell University Press.

Viguers, Stephanie, and Audrey R. Odom John. 2018. "Trained Rats More Successful in Diagnosing TB than Standard Tests." *Infectious Diseases in Children* 31, no. 5: 18.

Wald, Jonathan. 2022. "Horror." *Environmental Humanities* 14, no. 2: 367–70.

Wallace, Julia. 2014. "Workers of the World, Faint." *New York Times*, January 18, 2014.

Wetzel, Corryn. 2022. "Backpack-Wearing Rats Could Start Search-and-Rescue Missions Next Year." *New Scientist*, June 17, 2022. https://www.newscientist .com/article/2324994-backpack-wearing-rats-could-start-search-and-rescue-missions-next-year.

Widyono, Benny. 2007. *Dancing in Shadows: Sihanouk, the Khmer Rouge, and the United Nations in Cambodia*. Rowman & Littlefield.

Wikan, Unni. 1990. *Managing Turbulent Hearts: A Balinese Formula for Living*. University of Chicago Press.

Wool, Zoë H. 2015. *After War: The Weight of Life at Walter Reed*. Duke University Press.

Work, Courtney. 2017. "The Persistent Presence of Cambodian Spirits: Contemporary Knowledge Production in Cambodia." In *The Handbook of Contemporary Cambodia*, edited by Katherine Brickell and Simon Springer, 389–98. Routledge.

Yin, Cheryl H. 2020. "Khmer Has No Grammar Rules: Metapragmatic Commentaries and Linguistic Anxiety in Cambodia." *Journal of Southeast Asian Linguistics Society* 6: 83–111.

Zani, Leah. 2015. "Bomb Ecologies? Inhabiting Disability in Postconflict Laos." *Somatosphere*, Inhabitable Worlds Series. http://somatosphere.net/series /inhabitable-worlds.

———. 2019. *Bomb Children: Life in the Former Battlefields of Laos*. Duke University Press.

Zucker, Eve. 2013. *Forest of Struggle: Moralities of Remembrance in Upland Cambodia*. University of Hawai'i Press.

Index

Founded in 1893,
UNIVERSITY OF CALIFORNIA PRESS
publishes bold, progressive books and journals
on topics in the arts, humanities, social sciences,
and natural sciences—with a focus on social
justice issues—that inspire thought and action
among readers worldwide.

The UC PRESS FOUNDATION
raises funds to uphold the press's vital role
as an independent, nonprofit publisher, and
receives philanthropic support from a wide
range of individuals and institutions—and from
committed readers like you. To learn more, visit
ucpress.edu/supportus.

www.ingramcontent.com/pod-product-compliance
Lightning Source LLC
Chambersburg PA
CBHW030843270326

41928CB00007B/1200